PRACTICAL GEOGRAPHY

Practical Geography

A.M. Bagulia

ANMOL PUBLICATIONS PVT. LTD.
NEW DELHI - 110 002 (INDIA)

ANMOL PUBLICATIONS PVT. LTD.
H.O.: 4374/4B, Ansari Road, Darya Ganj,
New Delhi-110 002 (India)
Ph.: 23278000, 23261597
B.O.: No. 1015, Ist Main Road, BSK IIIrd Stage
IIIrd Phase, IIIrd Block,
Bangalore - 560 085 (India)
Visit us at: www.anmolpublications.com

Practical Geography

First Published, 2006

ISBN 81-261-2845-3

PRINTED IN INDIA

Printed at Mehra Offset Press, Delhi.

Contents

Preface

The study of earth and all that relate to it in terms of flora and fauna plus all that pertain to its inside-outside is called Geography. Among various subjects of study, the formation and structure or shape of earth is of the greatest interest.

Since the 'Big Explosion', popularly known as the 'Big Bang', theories have come in thousands of numbers, talking about the shape and various aspects of the most opulent and rich planet, with reference to life-supporting resources of the solar system. But the most approved-of concept, regarding the shape of the earth is that it is ball-like.

Metaphorically, this 'Big Ball' has multiple layers in its bosom, very similar to an onion. These strata are called: crust (floor or surface), upper mantle, lower mantle, outer core and inner core.

Born approximately forty five crore years back, our beloved earth has an area of fifty crore and ninety six lakh square kilometres. Having a 4500°C temperature at its inner core, its outermost crust is pervaded by a number of gases including oxygen with the maximum proportion of 46.6% —a life-line.

Of all these important facts, we are here, exclusively concerned with the part, forming its outermost layer. Divided between the two, the facial stratum comprises land and water.

For ease, this land area has further been divided into seven regions, called continents. They are : (1) Asia, (2) Europe, (3) Africa, (4) North America, (5) South America, (6) Oceania, (7) Antarctica.

That way, the part of earth, we belong to, is Asia. Furthermore, India falls under South Asia, one of the most sensitive and interest-gravitating sphere in the global geography.

Public and private libraries possess stacks of books and encyclopaedia on this subject though, the undersigned has been perpetually under the weight of hovering ideas and thoughts of such a compact, exclusive and comprehensive work as may address any issues, whatsoever in connection with Geography.

Here is that unique book, nay, a series of semi-text books in your service for your ready reference. Knit neatly and prepared with every possible precaution to not let any minor error creep into its fabrication, this elegant piece of work includes five books, dealing with : (1) Physical Geography, (2) Economic Geography, (3) World Geography, (4) Indian Geography, (5) Practical Geography.

I feel proud of dedicating this beautiful series to my beloved motherland, India.

Henceforward, begins your duty; cherish and relish it word-by-word and acknowledge me with your honest feedback.

— Editor

1

Concepts and Perception

Indian society is characterized by glaring diversities in its social and cultural features. The roots of these diversities lie in the hoary past. Tracing the origins of social differentiation, in itself a challenging task, may be necessary in order to understand the pluralistic nature of the present-day society.

Locational Setting : From time immemorial different streams of cultures originating from western, eastern and central Asia have made incursions into the South Asian sub-continent. Being at different stages of social evolution these culture-groups found for themselves ecological niches which suited the continuation of their specific modes of living. A keen observation of the social scene in India today may help us in perceiving a broader relationship between the way the people of India passed through different stages of social evolution and the emergence of diversities in levels of cultural and social development. In the course of this chequered social history these primeval communities were distributed and redistributed in space under different situations promoting in some areas cultural intermingling and in others isolation in geographical space. Early incursions were made from different entry points in the Hindu Kush Himalayas in the west, northwest, north and northeast, making the northern region of the Indian sub-continent a recipient of a variety of culture-groups.

The physical framework of the sub-continent determined their initial routes of migration as well as their choice to remain confined to the river valleys or the inter fluvial tracts as the case may be. These groups wandered freely in space over vast tracts of land, with varied ecologies which were either inviting or repulsing. Their freedom of movement was curbed by the scarcity of food as also by the natural barriers, such as rivers, hill ranges or the desert expanses. Over the wide open valleys and the undulating plains on the plateau, they could also perceive social barriers along the contact lines between different ethnic groups; and the barriers were enforced by social custom.

It may be assumed that the primeval culture domains were ethnically homogeneous and socially defined, as their violation invited conflict, and internecine wars followed. Initialiy, there was hardly any possibility of cultural overlap. Subbarao demonstrated with the help of massive archaeological evidence how this state of mutual exclusivity has best been preserved in the remote hilly or the mountainous tracts, which were characterized by a certain degree of geographical inaccessibility. Those occupying the open river valleys from Punjab to the deltaic region of Bengal were always under pressure, as fresh incursions were made, resulting in their ouster and eventual exodus from the scene of conflict. These groups in their own turn moved into the bordering rimlands, characterized by hilly terrain, desert or forest landscapes.

Historical Background

Indian history is full of references to such episodes in which incursions were made resulting in the ouster of the groups who came earlier and their eventual migration to the interior areas, the so-called refuge zones. Each incursion marking the arrival of a fresh cultural wave ended in the displacement of an earlier culture-group which was eventually pushed into the *cul-de-sac*. These incoming culture streams left behind a pattern in space with the earliest culture-groups to be found in the isolated and the remote pockets, farthest from the entry points in the northwest, and the youngest and the newer elements in this ever-changing cultural scenario, in the fringe areas placed closest to the entry points. Those who managed to drift into the isolated and the remote parts of the sub-continent could preserve their original ethnic traits which remained by and large unaffected by the fresh waves of incursions regularly witnessed in and along the main corridor of movement connecting the Kabul valley with the North Indian Plain. We may trace the origins of social differentiation and link it to this process of incoming cultural streams at different stages of social evolution and their subsequent isolation and placement in different segments of geographical space.

The physical framework—relief features, altitude, climate and more importantly, water supply— supported in a neatly defined manner the scope for the plurality of forms of culture, with the new elements co-existing with the old, the archaic and the primitive simultaneously, and surviving over time. While the newest elements were assimilated by the groups living all along the main route of migration, the continuity of the old, the primitive and the archaic elements distinguished the isolated regions in the backyard of this arena. One cannot dismiss this historical evidence easily. The evidence suggests a tiered pattern in the spatial arrangement of cultures. Take, for example, the north-western region of India, consisting of sharply defined eco-regions, such as the valley of Kabul in Afghanistan, the plains of Punjab from Indus to Yamuna, the upper valleys of the Punjab rivers, such as the Satluj and Beas valleys in Himachal Pradesh, or the upper valleys of Chenab in Kishtwar and the

Jhelum valley in Kashmir, and the trans-Himalayan regions in Ladakh, Baltistan, Gilgit and Hunza.

The historical experiences of the people in these sharply contrasted eco-regions have been strikingly different. The process of *cultural renewal* has perhaps been most active in the sprawling plains from Indus to Yamuna. Social groups in the upper valleys (*e.g.*, Chenab, Sutlej, Beas) lived at a level of equilibrium with Nature which allowed the continuity of older traits over longer periods of time.

The valley of Kashmir offered yet another example of an agrarian culture persisting over time, at a low level of technology unable to advance beyond a certain level. The surrounding rim of hills and mountains accommodated even more primitive tribal elements at a safe distance from the impact of river valley civilizations. The restless Pathans, the transhumant Gujjar-Bakarwals and the Baltis, Ladakhis, the Lahaulis and the Kannauris lived in their own protected worlds in a state of stagnation, the passage of time leaving little imprint on their mode of living or fund of recycled ideas.

A social geographical interpretation of this phenomenon is understandably based on an intimate knowledge of the historical processes which unfolded themselves over the millennia. Eminent archaeologists, such as F.J. Richards, Mortimer Wheeler, Fairservis, Subbarao, Allchin and Sankalia, historians such as Toynbee, K.M. Pannikar and D.D. Kosambi and geographers such as O.H.K. Spate noted a fundamental relationship between the physical and cultural regions in the Indian sub-continent.

Indian System

The Indian sub-continent marks a southward extension of the great landmass of Asia. The peninsula tapers towards the Indian Ocean as the two flanking expanses of the sea—the Bay of Bengal and the Arabian Sea—define its long winding coastline. Before the advent of the sailing boat or the steamship, which was a much later development, the sea must have played an important barrier role. In fact, the oceanic realm kept the Indian peninsula isolated from the littoral regions in Africa, West Asia and South-east Asia. It is this isolation promoted by the oceanic realm which appears to have contributed to the uniqueness of the Indian civilization as it evolved through time. Likewise, the landmass on the north is girdled by a chain of high mountains and inter-montane plateaux from west to east with the mountains forming a wall-like barrier between North India and the trans-Himalayan regions in Central Asia and West Asia.

The mountains have thus acted as a formidable barrier making all approaches to India difficult. However, it does not mean that the mountains were impregnable. More enterprising nomadic groups continued to wander through the high altitude passes in the Hindu Kush Himalayas. In fact, these passes remain snow-bound during the greater part of the year and can be negotiated with great difficulty during the short summer period.

It is during these periods that the nomadic groups moved with their herds of animals and crossed the Hindu Kush-Himalayan chain. Thus, the mountains on the north and the Indian Ocean on the south have both underlined the partially enclosed character of the Indian sub-continent. It is this geographical isolation of a certain degree which has contributed significantly towards the cultural homogeneity within the South Asian region .

With the passage of time the South Asian sub-continent evolved as a distinct cultural region, differentiable unmistakably from the West Asian, Central Asian and South-east Asian cultural realms. Girdling the sub-continent on the north, the Himalayas stretch in a northwest-southeast direction for a formidable length of 2,400 km between the Indus gorge in the north-west to the Brahmaputra gorge in the north-east. The Himalayas, however, are not a single chain of mountains. The mountains actually consist of three main series of ranges running parallel to each other from south to north. They are often referred to as the Sub-Himalayas, Lesser Himalayas and the Greater Himalayas or as the Outer, Middle and Inner Himalayas. These ranges, which have varied widths and differ strikingly in structure and relief features, are separated by intervening, longitudinal valleys of tectonic origin and the inter-montane plateaux, being the residual erosion surfaces of the earlier cycles of denudation. The mighty streams that cut across the Himalayas such as the Indus, the Sutlej and the Brahmaputra, appear to have given the Himalayas the contours of their commonly perceived transverse sectoral divisions—Punjab Himalayas between the Indus and the Sutlej, Kumaun Himalayas between the Sutlej and the Kali, Nepal Himalayas between the Kali and the Tista and Assam Himalayas between the Tista and the Brahmaputra (Dihang).

To the north of the Himalayas the terrain is dominated by diversely arranged chains of mountains in the northwest, *viz.*, Karakoram, Hindu Kush, Kun Lun, Tien Shan, Pamir-Alay and Trans-Alay ranges, which converge on the Pamir promontory. The Himalayas merge into these ranges imperceptibly giving the relief a complexity of a very high order. In Kashmir, the Great Himalayan range is skirted by the Ladakh and the Zaskar ranges. Further eastward they coalesce with the Kailash range which runs parallel to the Himalayas, on the outskirts of the Tibetan plateau. The surface features are further complicated by the presence of earlier erosion surfaces, such as Aksai Chin, Deosai, Depsang and the Lingzi Tang. Like Dehra Dun and Patli Dun, the Kashmir valley in itself is of tectonic origin—an end-product of the differential uplift during the Himalayan orogenic phase. These narrow longitudinal valleys and the river terraces bordering them, particularly the Kashmir valley, became centres of convergence of early human cultures.

Despite the formidable nature of the mountain ramparts, there are passes all along the length which have enabled the movement of adventurous nomadic groups throughout human history. This is notwithstanding the fact that most of these passes in the north, northwest and the northeast are situated at high altitudes and are beset with great difficulties. Understandably, trans-Himalayan communication has been an adventure always. Some of these passes are well-known, *e.g.*, Gomal, Khyber and Bolan in the Sulaiman range. Khyber has served as a gateway to India throughout history. However, the pass is accessible from the Kabul valley route and once the Bamian pass in the Hindu Kush range has been crossed it is the only natural entry point.

Other high altitude passes which have served as gateways to India include Karakoram, Zojila and Rohtang in Himachal Pradesh. There are passes over the Barahoti plateau in Uttar Pradesh, in Nepal Himalayas, in the Chumbi valley in Sikkim and in Arunachal Pradesh as well as within the Indo-Myanmar hills. The river gorges, such as Dihang, in the northeast, have also served as corridors for incursion into India. The same role has been played by the numerous streams in Baluchistan connecting the Indus region with Iran.

The Afrasian dry zone—the river valleys of West Asia *(viz.,* Tigris and Eupherates) and North Africa, *(viz.,* Nile)—has been recognized by historians as "the most important region at the dawn of civilization". These steppe lands of western Asia and the dry zone of North Africa including the rim of the Mediterranean became eminent centres of human dispersal. These regions were fragile from the viewpoint of human survival. Early Humans who appeared in this area faced the chronic problem of adjustment with the natural environment which was characterized by both the scarcity of water and grassland resources. Moreover, historically the region suffered from increasing aridity.

This 'progressive desiccation' is believed to have resulted in a continual process of radial migration of people to the east. The streams of migration brought successive human cultures to the Iranian plateau, as well as the regions lying to the south of the Caspian Sea and the Indo-Iranian borderlands in Baluchistan. By virtue of its location on the terminal end of the Eurasian landmass, India obviously became a focal point for some of the migration streams.

The Indus valley and the northwestern region of Punjab as they lay closer to the entry points became centres of attraction on a perennial basis. The traffic was, however, restricted by the formidable nature of terrain dominated by the Hindu Kush-Himalayan mountain chain.

The steppe lands of East Asia (*viz.*, Gobi desert and northern China) emerged in history as another centre of radial migration. From here the nomads drifted to the west taking trans-continental migration route lying to the north of the Central Asian mountain complex. The migration streams originating in North China and Gobi eventually converged on the valleys of Oxus (Amu Darya) and Jagxartes (Syr Darya), south and east of the Caspian Sea. These migrations often referred to as nomadic outbursts threatened the very base of the little oasis states that emerged in the river valleys separated by the great deserts of Kara Kum and Kyzyl Kum. Thus Central Asia remained an area of instability throughout history.

As compared to this India enjoyed a certain degree of stability rendered by its location protected by the high Hindu Kush Himalayan mountain massif and the Karakoram and Tien Shan ranges. The transcontinental migration route connecting North China with the East European Plain remained a major line of thrust for the nomadic traffic. Somewhere in the Amu-Darya Syr-Darya interfluve, a route bifurcated from the main line of thrust. It became possible to reach the northern face of the Hindu Kush by this route. Here an high altitude opening—the Bamian Pass—was taken by wave after wave of migrating groups to reach Afghanistan and on way to India.

As an entry is made through the Bamian Pass, the natural route opens up into the Kabul valley, eventually converging on where now the cities of Kabul and Jalalabad are situated. The Kabul river flows eastward, past these old nodes of power to meet the Indus near Attock, the Khyber pass offering an easy entry into the region. Toynbee has noted that throughout history all crossings of the Indus have been made at this strategic point. It is near here that the Indus debouches from the hills, and is no more a turbulent mountain torrent.

Downstream the channel of the river widens with the result that fording becomes an absolute necessity. Over time Attock acquired its nodal character emerging as the most important strategic point for a passage to India. From Attock traditional routes which follow a line parallel to the foothills of the Himalayas through the Punjab plains lead to Ambala-Kurukshetra region.

Southern routes were beset with many difficulties, arid climate and scarcity of water being most formidable among them. From Ambala it is possible to drift on to Kurukshetra Thaneshwar (Sthaneshwara of Huen Tsang) and to Mathura via Indraprastha on the west bank of the Yamuna (Jumna). There has been a strong tendency to cross the Yamuna at a point where it is closer to the hills near Yamunanagar to reach the fertile region of the doab. A crossing of the Yamuna at this point makes the Ganga-Yamuna doab easily accessible from the west.

The natural routes then run parallel to the Ganga. The historic cities of Hastinapura, Kanyakubja and Prayaga arose along this route. Likewise, a crossing of the Yamuna near Mathura makes possible an entry into the doab and a passage to Kanyakubja. In this sector one is more likely to take a route parallel to the Yamuna. Historic cities, such as Chandwar,

Rapri and Kosambi arose along this route. Eastwards beyond Kashi the routes end up into the Ganga-Sone triangle. It was in this region that Rajagriha and Patliputra emerged as nodes of Magadhan power at different points of time in history.

The outliers of Chhotanagpur plateau in Mukameh spurs and the Rajmahal hills abut the Ganga forcing the river to take an abrupt southerly bend, thus defining the partially enclosed character of the south Bihar plain. Eastward movement into Bengal was beset with difficulties partly due to the highly dissected terrain. The Rajmahal hills thus emerge as a terminal point where the North Indian plain gets lost as an ecumene. The mighty Ganga on the north and the Chhotanagpur plateau on the south seem to have contributed to the region's psyche of autonomy and self-sufficiency. Astride this route from Indus to Rajmahal hills lie the historic cultural nodes from Sapta-Sindhava and Kurukshetra to Panchala, Braj, Kosala and Magadha.

From here the movement towards the south became possible via routes which passed through the Sone-Narmada furrow. These routes finally open up into the Malwa plateau and converge on the old city of Ujjaini. Narmada valley route takes one further to Gujarat and the Desh region of Maharashtra. Here, the upper tributary streams of the Godavari (*e.g.*, Purna) offer convenient routes to Telangana and the littoral Andhra where the deltas of Godavari and Krishna coalesce to form an alluvial tract of great agricultural potential along the coast.

The coastal plain widens here and the coastal route takes to the south upto the Kaveri delta. The Kaveri valley route takes one towards the Western Ghats and finally ends up in Malabar. Along this route from Ujjaini to Madurai (from Narmada to Kaveri) emerged the nodes of regions, such as Gujarat, Maharashtra, Telangana, Andhra, Chola, Pandya and Kerala.

Regional Structure

It is commonly known that India is a mosaic of regions. The river valleys from Indus to Kaveri have been the areas of perennial attraction, particularly for the agriculturally oriented communities. On the outskirts of these valleys lie the areas of rugged terrain consisting of hills and dissected plateaux not much suited to the agricultural mode of production. These regions have been less hospitable and more favoured by groups engaged in hunting, food-gathering or pastoral practices. These diversities in the ecological setting and the corresponding differences in human adaptation created diverse regional patterns in culture which are a distinctive trait of India.

Developing the theme of unity in diversity in the Indian sub-continent, Subbarao noted that 'significant diversities' in the different regions of the country suggested a more fundamental relationship between the physical and cultural elements which defined India's regions. In his attempt to understand Indian history and archaeology in better light, he tried to identify and analyze factors which operated behind the baffling cultural diversity. The geographic factors have helped to a certain extent in developing the understanding of the underlying cultural processes. However, in relating culture with environment one cannot overemphasize physical factors and thus follow a deterministic line. Human relationship with Nature is a complex process and the human failure to bring about homogeneous cultural (social) development cannot be attributed entirely to the environmental factors.

F.J. Richard's pioneering contribution to the interpretation of Indian history appears to have influenced archaeologists, historians and geographers alike. Spate in his reconstruction of nuclear areas from Asoka to Aurangzeb, revealing frequency of political boundaries as they evolved through time, was trying to interpret Indian history in the geographical context (Box).

> ***The Perennial Nuclear Regions :*** Thus early we can also discern the emergence of some nuclear regions or bases of power which are perennially significant in Indian historical geography: Gandhara in the Vale of Peshawar and Potwar, Sapta-Sindhu narrowed down to the Punjab, seven rivers to five; Kurukshetra (Sirhind), the Delhi or Sutlej/Yamuna Doab; Panchala in the Yamuna/Ganga Doab and Rohilkhand; Saurashtra (Kathiawad) and Gujarat; the four great kingdoms already apparent in Magadhan times. In the Dravidian south the pattern is more confused, but not without some relatively permanent pieces in the dynastic kaleidoscope: the Kalinga country or Orissa; Andhra, the Telugu country; the Chola (whence Coromandel) and Pandya kingdoms in the Tamil country; Kerala or Malabar, the isolated southwest littoral. There are of course many smaller areas which have preserved a historic individuality, *e.g.*, Bundelkhand, Chhattisgarh, Konkan and Kanara. Some areas again have been debatable marches: such are Khandesh, between Narmada and Tapti, or the Raichur Doab between Tungabhadra and Krishna.
>
> —O.H.K. Spate, *India and Pakistan*, 1957: 177

The foregoing study of the space relations of India with the adjoining regions of Asia has indicated the directions from where the cultural influences percolated into the sub-continent and affected the course of historical development. The impregnable nature of the northern mountain complex (Hindu Kush Himalayas) and the location of the sub-continent as a peninsula projecting into the Indian Ocean protected the region from the upheavals that affected the Central and West Asian regions. Whatever cultural influences managed to reach India, they were in the form of a 'backwash' from the main line of thrust that lay to the north of the Tien Shan range in Central Asia.

While the early man *(homo sapiens)* is not traceable in India, it is widely known that "in the subsequent periods... every new wave of people, who entered the Indo-Gangetic plains as conquerors or fugitives very rapidly lost their own individuality in this melting pot of cultures and added their own indistinguishable element to the makeup of this culture complex". Some of the dominant strands left their 'deep imprints' while others remained 'unidentified and unaccounted.

The division of the country into 'wide inviting alluvial plains' connected to the principal gateways opening into western Asia and the southern peninsular plateau with the intermediary stretches studded with hills and the tropical monsoon forest, provided ideal ecological conditions for the survival of a variety of cultural traits, enabling "the earlier inhabitants to isolate themselves in what may be called *cul-de-sac*, or refuge zones". The geography of India contributed to this spatial arrangement. As already pointed out Richards and Subbarao noted the role of relief features in the horizontal expansion and segregation of cultures.

"The chief river basins of the country", wrote Subbarao, viz., "the Indus, Ganges, Narmada, Tapi, Godavari, Krishna and Kaveri, were in turn penetrated and exploited by large scale agricultural communities, driving the older and static people in a more primitive economy into the forested mountains, where they have survived to this day. The main river basins of the country with a rainfall between 20 and 40 inches (500-1,000 mm), which can sustain large scale agricultural communities, have been colonized or occupied again and again."

India has evolved in its spatial segments a typology of regions, described as (a) areas of attraction or perennial nuclear regions; (b) areas of isolation or cul-de-sac; and (c) areas of relative isolation lying in between. This scheme of regions is closely linked to the 'trans-continental communication system' through which people and cultures have moved from the northwestern entry points to the southern tip of the peninsula (Box).

> *The Perennial Nuclear Regions* : Now working on this system, we can easily define the regions and their foci. Bounded by the Aravallis and the desert of Rajputana in the east, and the Sulaiman and Kirthar ranges in the west, is the Indus basin, draining the central Himalayas. This can be roughly subdivided at the point where the hills from the west and the desert in the east converge near the Bugti country, into the lower and the upper, corresponding to Sind and Punjab respectively. Beginning with Delhi, where the Aravallis converge towards the Himalayas, the Gangetic basin runs east from the narrow divide. On its southern flank lies the Vindhyan complex. The lower deltaic region of the Ganges, as it leaves the Vindhyan complex, is Bengal. The valley of the Brahmaputra is Assam. Forming, as it were, a little triangle constituted by the Aravallis in the west, and the Vindhyas running obliquely towards the Gangetic basin, is the plateau of Malwa, drained by the rivers—Chambal, Banas, Sipra, Narmada and Son. Lying south of the Vindhyas, and constituting the upper basins of the rivers Krishna and Godavari, and more or less co-extensive with the Deccan Trap area, is Maharashtra. The lower basins of these two great rivers constitute Andhra. The southern part of the Krishna basin, more or less constituting the 'rocky triangle' formed by the Eastern and the Western Ghats, *viz.*, the Mysore plateau, is Karnataka. At its southern end, it is drained by the river Kaveri.

> Beginning from the constriction of the Nagar hills and Pulicat lake, and running right along the east coast is Tamilnadu. The narrow coastal plain on either side of the Palghat gap in the Western Ghats is Kerala. The valley of the Mahanadi, surrounded by hills and plateau, constitutes Orissa. The northern part of the west coast and the peninsula of Saurashtra, abutting on the desert of Rajputana, is Gujarat.
>
> —B. Subbarao, *The Personality of India*, 1958: 16-18

According to this viewpoint, the difference in the cultural milieu of the first large scale agricultural communities in different regions contributed to the baffling diversity that India inherited. The movement of people and cultures through the trans-continental communication system promoted intermingling of cultural traits thus ensuring the over-arching cultural unity within the diversity. Expansion of the primeval agricultural communities from the core areas of the river basins has resulted in an encroachment on the tribal cultures surviving in the surrounding rim of hills and forested tracts in the uplands. This has initiated a process of cultural assimilation or absorption of tribal traits into the dominant peasant cultures.

Its natural consequences were disturbing to the tribes whose fragile economies were destabilized with these peasant inroads and could never recover from this shock. The juxtaposition of cultures observed in the regions surrounding the river basins, in the Chhotanagpur plateau, Orissa highlands, Central Vindhyan Complex, the Aravallis or the Western Ghats, is an example of this uneasy interface between the tribal and peasant modes of production in contemporary India.

It is thus evident that the geographical setting of India played a very significant role in the evolution of cultural diversity.

The North-east Region : Like the rest of India the northeast also passed through the same evolutionary history. The same process of incursion of diverse cultures, their horizontal expansion and consequent displacement, contraction or isolation of earlier cultures also affected the northeast. The region is often described as a microcosm of India with its northern mountains, central plain and southern plateau.

The Brahmaputra has filled its tongue-like longitudinal valley with a massive alluvial deposit of great agricultural potential. However, the valley remained a rendezvous for hunters and food-gatherers far longer and the process of sedentarization started quite late in history. The evidence of shouldered celts (stone hoes) scattered all over the region suggests beyond any doubt that the earlier culture-groups were familiar with the art of agriculture. However, the process of sedentarization was postponed by many centuries.

The setting was ideal for the incursion of diverse cultures and for their eventual segregation in the Brahmaputra valley and the surrounding uplands consisting of Garo, Khasi, Jaintia, Lushai and Naga hills. The land routes connected the Brahmaputra valley with the hills and plateau surfaces beyond Tibetan plateau, Shan plateau and the Irrawaddy valley. The Chittagong hills also remained connected with the Cape Negaris and the Malay peninsula through the Arakan coastal route.

Moreover, the numerous passes across the Himalayan range and the Indo-Myanmar hills offered access to the Brahmaputra valley which became an area of convergence of diverse cultures. The Ahoms who came to the region during the thirteenth century transformed the scene completely resulting in the division of the agrarian and the pre-agrarian modes of production along topographical lines. Today, the arrangement is such that while the tribes are more or less confined to the hills, the valley is the home of peasant cultures. Manipur replicates the same arrangement on a much smaller scale.

The Brahmaputra valley emerged as the far-eastern

perennial nuclear region with its central and western segments known variantly as Kamarup or Pragjyotisa. For the simple reason that the western region was geographically isolated from the rest of northern India, it enjoyed a certain amount of independence throughout history. Even though the spill-over from the densely populated Ganga delta to Assam remained a phenomenon, particularly after the Permanent Settlement during the British period. Even the Muslim kings from Khilji to Akbar could not control the North-east in its entirety. The region was eventually annexed by the British during the nineteenth century who set the stage for the new political order and a more comprehensive colonial exploitation of natural resources.

Diverse Modes and Spatial Arrangements : It may be recapitulated that the river valleys were more attractive to the agrarian groups. The emergence of nuclear regions was obviously a post-agricultural phenomenon. However, before the advent of agriculture, hunters and food-gatherers were not interested in the river valleys for any particular reason. They wandered from vale to vale in quest for food. India presented a diversity in its ecological setting to such an extent that the tribal, agrarian and pastoral modes could survive in a neat spatial arrangement. While the agrarian mode spread horizontally in the river valleys, the inter-fluvial tracts were favoured by the nomadic pastoralists. Higher up on the plateau surfaces, or in isolated hills and mountain ranges tribes lived in their own exclusive world.

Regional Nuclei : As already noted, the nuclei of perennial regions were in the river valleys, big and small, from the Indus to Krishna and Kaveri. The underlying assumption is that the river valleys were accessible by the main route of sub-continental migration. The route inter-connected them and kept them open to fresh waves of incursions time and again. Along this route emerged Gandhara (west of Indus at the end of the Kabul valley), Sapta-Sindhu (upper Indus basin),

Kurukshetra (between Sutlej and Yamuna), Panchala (between Yamuna and Ganga in the upper doab as well as trans-Ganga region), Braj (along the Yamuna), Kosala or Awadh (north of the Ganga in its middle segment), Magadha (south Bihar plain between the Sone and the Ganga), Bengal over the Ganga delta (between Rajmahal hills and the sea) and Assam (in the lower Brahmaputra valley).

In Central India emerged Malwa (between the Aravallis and the Vindhyan range), Maharashtra (upper basins of Krishna and Godavari, south of the Vindhyas), Andhra (lower basins of Krishna and Godavari), Tamilnadu over the eastern littoral region and Kerala (west of the Western Ghats). Other nuclear regions may be identified as Orissa (lower Mahanadi basin) and Gujarat (basins of Mahi and Sabarmati as well as Kathiawar peninsula).

It may also be noted that while Orissa enjoyed a certain degree of isolation because of its location away from the main route of sub-continental migration, Gujarat remained connected to this circulation through the mid-Indian as well as the Makran coast routes. In fact, it remained in constant contact with West Asia by sea as well as land routes. Movement along the Makran coast route influenced the cultural history of Gujarat in several phases.

On the outskirts of the river valleys were the so-called refuge areas—hilly, forested or arid. Foremost among these *culs-de-sac* were Rajputana (west of the Aravallis), Gondwana (between Sone and Mahanadi), Bastar plateau in Central India and Rajmahal hills. Then there were sub-Himalayan river valleys of Chenab, Satluj and Beas in the old Punjab, as well as Kumaon and Garhwal hills. These regions served as refuge areas for both the tribal and peasant groups, particularly for the latter, who were ousted from the plains of North India after each successive wave of invasion. Scions of the displaced ruling class often found refuge in these remote hills from Dhauladhar to the Kathmandu valley founding their own little, semi-feudal, states in the Himalayan retreat.

Mention may also be made of Kashmir and Ladakh—regions nestled in the Himalayas and beyond. Kashmir, however, remained connected with Gandhara by the much frequented Jhelum gorge route, even though its 'refuge area' character remained unaffected. Ladakh was away from the Kashmiri world and continued to retain Buddhism in its original form, despite the advent of Islam in the valley of Kashmir and the bordering rim of mountains from Baltistan to Chitral.

In the northeast the bordering uplands from Arunachal Pradesh to Lushai and Chittagong hills served the role of refuge areas for the tribes ousted from the river valleys—the tribal-non-tribal boundary runs often parallel to the 'skirts of the hills'. Those who joined later such as the Kachins and Kuki-Chins, found no other place but the hills which were already partly occupied by earlier groups.

Evolution of Cultural Differentiation : The presence of the Early Humans in the Indian sub-continent is inferred from the tools of the Old Stone Age collected by the archaeologists from countless number of sites scattered all over the region. Even though these tools provide indisputable evidence of the existence of the Early Humans, no direct evidence has yet been unearthed from any site which may throw light on the very nature of this population, their racial characteristics and physical attributes. Thus, while a multitude of tools is there all around, human skeletal remains are missing. This issue has been widely debated by archaeologists and physical anthropologists. However, it may not be a very serious problem as the tools themselves are sufficient to tell a coherent story about the evolution of Early Human cultures from the Palaeolithic (Old Stone Age) to the Neolithic (New Stone Age) and the Chalcolithic (metal using cultures) periods. A socio-geographical interpretation of this evidence of the human artifacts belonging to the different pre-historic phases is important in order to understand the very process of human adaptation to the prevailing environmental conditions and the human migrations related to this adaptation.

The evidence also helps in the appraisal of emergence of certain localities which were repeatedly frequented by the Early Humans in the course of geographical spread of these cultures. In fact, it is customary to recognize three types of regions, each one of them throwing up evidence of diverse cultural history. Broadly speaking, while regions of attraction possess evidence of a complete cultural succession reflecting a remarkable continuity, the regions of isolation are different. They were not frequented so often. Their archaeological record is, therefore, discontinuous. Whatever cultural influences managed to reach there, they virtually got impounded as no new influences could percolate. In between these two extremes are the regions of relative isolation, where cultural influences were received but with breaks, thus leaving behind an archaeological record of a discontinuous nature.

The pre-historic scene in India is hazy because of the limited nature of evidence. The inadequacy of probings made into the past is an obvious explanation, although the past is a closed book unless its pages are opened systematically. This has left vast gaps in information which have often been filled with conjecture. In fact, many sites were accidentally discovered. The selection of sites for digging was also subjective. Moreover, archaeological excavations progressed in the light of assumptions, hypotheses and formulations arrived at similar diggings else where in the world.

The schools of European archaeology had their own models which were neatly applied to Indian data. Despite the random nature of diggings, the evidence was colossal, both in extent and content and called for rigorous analytical and interpretative exercises on a scale proportional to the size of the territory.

Understandably, some sites received more attention than others and at any point of time findings based on new diggings upset the very basis of earlier conclusions. There were, however, serious limitations, both administrative and financial, which made the archaeologists' task even more difficult. Reaching the goals set by the Archaeological Survey of India was always as difficult as it is today. The archaeological record is, therefore, to be seen with care and caution.

The earliest traces of human cultures have been found in the valley of Sohan (Soan in early writings) river, a tributary of the Indus, northwest of Rawalpindi (Pakistan). The evidence collected from the terraces of the Sohan river consists of crude stone tools believed to be left behind by the ancient hunters of the Old Stone Age. These tools were found buried in the glacial as well as the fluvial terraces of the Sohan river. The terraces also presented a complete stratigraphic record of the sequence of climatic changes witnessed during the Pleistocene times. Since the tools were found buried in the river terraces, there was a valuable stratigraphic evidence, enabling the archaeologists to determine the age of the tools with the help of strata with a certain degree of precision. In fact, it was due to this strati graphic association of the tools that Sohan valley sprang into prominence as one of the most celebrated sites of the lower Palaeolithic period. The Sohan valley technique of tool-making became a reference point in comparative archaeology for checking the antecedents of similar tools discovered in other parts of India.

It was due to the pioneering work of De Terra and Paterson that it became possible to place the whole set of evidence collected from the Sohan river terraces in its proper perspective. By correlating the human cultures with the sequence of climatic changes during the Pleistocene times, De Terra and Paterson set the stage for a meaningful study of the Palaeolithic cultures in India. The Sohan valley tools were found in four of the five terraces of the Sohan river. The lithic evidence was collected from the terraces of the second glacial, second inter-glacial and the third glacial phases. In terms of evolution of tool-making techniques, archaeological studies revealed that three distinctive stages could be identified. These stages were termed as pre-Sohan, Sohan and the late-Sohan. Each successive phase revealed a technological advancement on the previous one. Thus, while the pre-Sohan tools were based on flakes, the Sohan and the late-Sohan varieties consisted of core tools. In fact, the early Sohan group, which belonged to the second inter-glacial period, consisted mostly of choppers and other scraping tools made on flat truncated pebbles. Typologically, they were sub-divided into several groups ranging from Clactonian to Levalloisian.

Placing the evidence in its proper environmental context, De Terra and Paterson noted a direct relationship between the climatic changes and the rise of early human cultures. It may be postulated that the tools unearthed from the terraces must have been left behind by the Early Humans over the surface in the Sohan river valley. The downward movement of the glaciers and the fluvial drift that followed the melting of snow must have displaced the tools from their original sites. Perhaps the Potwar plateau became one of the earliest sites of the diffusion of the early human cultures in the Indo-Pakistan sub-continent. The advances and retreats of glaciers implied vast fluctuations in the global temperature regimes. This must have resulted in far-reaching changes in the earth's environment, particularly in its climate and the vegetation cover. The ancient hunters and food-gatherers must have remained on the move in the wake of these changes.

Sohan-type tools have also been found in other parts of North India. Of special interest are the sites in the terraces of the sub-Himalayan river valleys of Satluj, Beas and its tributary, Banganga, in Himachal Pradesh. Similar tools have also been found scattered on the banks of the Yamuna in Bundelkhand

and along the banks of the Chambal and its tributary, Banas, in eastern Rajasthan. The pebble-based core tools (chopping choppers and scrapers) appear to have a southern limit somewhere along the 22°N Latitude, with Mayurbhanj, in Orissa, being the southern-most point where these tools have been found. Archaeologists have also found similar tools in certain stretches of the Narmada valley. Despite the widespread occurrence of these tools, no definite geographical picture of the development of Palaeolithic cultures can be reconstructed. As is generally known, the river valleys of the North Indian Plain have not yielded any tools. If any tools were left behind by the ancient hunters, they must be lying buried under the thick mantle of alluvium deposited during the successive phases of fluvial activity.

The known evidence of the lower Palaeolithic period in Peninsular India tells a different story. Perhaps the Old Stone Age tools discovered in and around Madras (now Chennai) and other parts of peninsular India belong to a different tool-making tradition. For one thing, the tools are not based on pebbles and mostly consist of hand axes and cleavers, made on cores and flakes. These Madras-type tools representing the 'peninsular facies' of the Early Stone Age in India are Acheulean in topology and suggest an independent development over the region lying to the south of the Narmada. Some observers have noted similarities between the Madras hand-axes and the tools found on the East African plateau, suggesting a possible connection.The Madras-type tools have been found in a vast area encompassing the peninsular region. Madras is recognized as a type area where in a boulder-conglomerate bed at the site of Vadamadurai, the earliest tools were unearthed. Studies conducted by Krishnaswami and Paterson revealed that there was geological evidence to suggest that the peninsular region experienced an alternation of wet and dry periods during the Pleistocene. These phases, which were termed as pluvial and inter-pluvial, left their traces in the deposits. While a direct connection with the glacial and inter-glacial cycles of Northwestern India has not been established, attempts have been made to correlate the stratigraphic and lithic evidence of the Madras region with the mid-Pleistocene fossil beds of Narmada. Other sites yielding Madras-type tools are situated in the terraces of the river Attirampakkam near Madras and the Krishna, Godavari and Narmada basins. Tools were also collected from Nellore as well as a large number of sites along the east coast of India and the Gujarat region.

It is thus evident that the appearance of Palaeolithic hunters on the Indian scene is supported by an overwhelming archaeological evidence scattered over the peri-glacial areas in the north-western Himalayas as well as the peninsular plateau. However, the Old Stone Age sites do not lend themselves to a systematic geographical interpretation, mainly due to the incomplete nature of our knowledge. The evidence also suggests that from the upper Palaeolithic to the Neolithic there was a transitional phase which is generally referred to as Mesolithic. During the Mesolithic period, one comes across an overwhelming evidence of smaller tools known as microliths. A multitude of these tools has been collected from sites scattered all over the sub- continent. The microliths include blades, scrapers, burins, triangles, crescents and lunates. The Mesolithic period marks a change from flakes to the microlithic tradition. The tools are generally made of finer varieties of rocket, such as jasper, chert, agate and quartzite. Also included in the findings are composite tools, made of stones fixed with bones or wood, as well as remains of pottery. These new varieties of tools indicate sophistication in the techniques of tool making, an imperative of the multiplying human needs. Insofar as the geography of the Mesolithic sites is concerned, nothing can be said with certainty.

In fact, each site was an isolated discovery and represented a type in itself. But these sites are important since they tell us the story of cultural succession from the Palaeolithic to the Neolithic. These tools are widely distributed in India, although no evidence has yet been collected from either the Ganga valley or the Himalayas. Sites have been extensively studied in western, central and southern India. In western India, tools have been found in the Mahi and the Banas river valleys of Gujarat and parts of Saurashtra, Mehesana (e.g., Langhnaj), Malwa and parts of Rajasthan. A layer at Langhnaj has yielded black, red and red-and-black potsherds.

Copper and iron tools have also been found at higher levels. Sites are also situated in Maharashtra, particularly in the upper Godavari valley (*e.g.*, Nevasa) and the Salsette island and Ellora. Microliths have also been collected from sites in Singrauli basin (*e.g.*, Lekhania rock shelters in Mirzapur district). Similar tools have also been discovered from the sites in the Damodar valley (*e.g.*, Birbhan-pur), Chaibasa, Chakradharpur, Santhal Parganas, Durgapur, Bankura, Singhbhum, Keonjhar, Mayurbhanj and Sambhalpur. Other rock shelters, such as the one at Adamgarh hill in the Narmada valley, are rich in the cultural remains of the Mesolithic period. Several sites have been excavated in parts of southern India, particularly in Chitradurga, Kurnool and Bellary districts (*e.g.*, Brahrnagiri). Not all the sites reveal a cultural continuity. For example, sites from where the hand axes were collected do not reveal a cultural succession to the Mesolithic. On the other hand, the *teri* sites in southern Tamil Nadu have yielded evidence of transition from the Palaeolithic to the Neolithic cultures.

It may be concluded that the evidence of the microliths points up to a gradual transition of the Palaeolithic hunters and food gatherers to an agriculture-based production economy. The transition from food gathering to food production was indeed a slow process involving millennia of years. Interestingly, some of the ancient hunters continued to remain with the Palaeolithic mode till very late in history. For them agriculture remained a peripheral activity, if at all. But in regions in which early crop raising became possible, the emergence and the spread of agrarian mode of production is evident. The agrarian mode eventually transformed the human society. This phase of cultural development is often referred to as Neolithic revolution. It involved a process of settling down of nomadic groups on pioneer lands reclaimed from the forest and adopting agriculture as their economic mainstay.

Neolithic Period

Insofar as the South Asian sub-continent is concerned, transition towards the Neolithic appears to be somewhat confusing. One general conclusion is that outside the Indus basin, the Neolithic cultures are marked by their late appearance—so late that they are mostly post-Harappan. This would mean that while the Neolithic cultures emerged in Baluchistan several millennia before, and, not only that, a civilization took its roots in the Indus valley and the adjoining areas in Gujarat as well as in the Ghaghar (Saraswati) valley in northern Rajasthan, the rest of the sub-continent remained in the Palaeolithic age. However, some sites in other parts of India have yielded evidence of Neolithic cultures which were certainly pre-Harappan. In fact, archaeological evidence suggests that the Neolithic and the Chalcolithic cultures overlap at many places and not necessarily reveal succeeding cultural phases. Sites in Baluchistan, to which a reference will be made in the subsequent paragraphs, parts of southern Deccan and the valley of Kashmir have thrown up definitive evidence of the presence of an earlier cultural phase which was unmistakably Neolithic. These Neolithic communities, although largely dependent on the use of stone, domesticated animals and practised a primitive form of agriculture. They were leading a settled life. This phase is followed by another cultural phase in which copper and subsequently bronze is associated with stone tools, thus indicating the emergence of a Chalcolithic culture. Although, Baluchistan leads in the development of the Neolithic cultural phase, there are other regions in the sub-continent in which the use of metals was quite common.

Baluchistan : The earliest village sites representative of the Neolithic cultures in the Indo-Pakistan sub-continent were dug out from the tiny river valleys of Baluchistan. That the first Neolithic communities emerged on the Indo-Iranian borderlands suggests the possibility of eastward shift of the agricultural mode from West Asia to the Indian sub-continent. The oldest village sites in West Asia are located in Turkey, Palestine and Iraq. Chronologically, they are supposed to belong to the seventh or the eighth millennium B.C. The Baluchistan cultures are, however, much younger and are dated around the middle of the fifth millennium B.C. While the time lag in this eastward shift of the Neolithic cultures is evident, their rise in Baluchistan and in other parts of the Indus valley may be seen as a remarkable development.

In fact, traces of the Neolithic cultures have been found over a vast geographical area encompassing Baluchistan uplands, Indus basin and the adjoining areas in northern Rajasthan. Archaeologists agree that this Neolithic phase was pre-Harappan in its origin and must have been a major influence on the later development of the Harappan culture which developed a civilization. Archaeological work at many of the Harappan sites has shown the presence of lower layers yielding evidence of the pre-Harappan (Neolithic) cultures. Ironically, these lower levels at many of the Harappan sites have not yet been dug out because of physical difficulties (*e.g.*, high water-table) leaving the question of Harappan-pre-Harappan connection unresolved. It belongs to the realm of conjecture.

The tiny rivers in the Baluchistan hills, Zhob, Beji, Pishin Lora, Dasht, Kel and Lora Lai, to name only a few, in their neatly carved-out valleys, offered ideal conditions for the rise of primeval village settlements dependent on some form of agriculture coupled with rearing of animals. Water supply in these rivers was seasonal and permitted the sowing of crops at different levels along the river banks at a height of 1,200-1,600 metres above sea level. There are many sites from where the evidence of this cultural phase was collected. Archaeological evidence throws light on the nature of the artifacts as well as houses and pottery traditions. Notable among these sites are Kili Gul Mohammed, near Quetta, Rana Ghundai in the Lora Lai valley, Periano Ghundai in the Zhob valley, Damb Sadaat, Togau, Siah Damb, Anjira, Nal and Amri: Kots Diji, east of the Indus, and Mundigak in the south-eastern region of Afghanistan. The evidence collected from these sites shows how the food-producing mode sustained these communities. This mode eventually spread to other areas in the Indus valley and set the stage for the rise of a civilization, the first in Indian history.

In a way these were the pre-Harappans who laid the foundations of the Indian cultural tradition, typically representative of an agrarian society. Figurines of mother goddesses (*e.g.*, Zhob goddesses) and the *pipal* leaf motifs on pottery leave much to be inferred about the origins of Indian civilization.

Harappan Civilization : The sites of the Harappan civilization are scattered over a vast geographical area: from Rupar in the Punjab foothills to Lothal and Rangpur on the Gulf of Cambay and from the Makran coast to Kalibangan on the dry bed of the Ghaggar in northern Rajasthan. Alamgirpur, east of Yamuna, appears to be the eastern-most outpost. The site of Bhagatrav lying on the Narmada extends the boundary of the civilization further south. It may be inferred that its impact was even wider than what the known sites presently indicate. More sites are being discovered and it is quite possible that the currently held view about its extent and boundaries is only tentative and may be revised as new evidence is put forth.

The transformation of a Neolithic culture with its modest beginnings in Kili Gul Muhammad, Rana Ghundai, Kot Diji and Kalibangan into a civilization with a substantial food-producing economy to support an urban society predominantly engaged in the artisan goods industry and in a widespread exchange of these goods on a commercial scale, must have been a time-consuming process. The evidence suggests that the civilization had already set in by the beginning of the third millennium B.C.

The fact that it could sustain itself for one full millennium and a half testifies on its ecological viability. The food-producing economy in a dry region with scanty rainfall could have been possible only in the event of rivers carrying sufficient supply of water to the agricultural fields. The Indus and its tributaries allowed the people to grow a variety of crops in the dry regions of Sind, Punjab and northern Rajasthan as these rivers carried with them a large volume of the snow-melt in the peak summer season. It appears that the surplus from agriculture was transferred to the numerous towns—Mohenjo Daro on the Indus, Harappa on the Ravi, Kalibangan on the Saraswati—to allow people necessary freedom to engage themselves in the manufacture of artisan goods. The raw materials for these industries came from far and near—a variety of minerals, particularly zinc and copper and precious stones from the Aravallis, cotton and other agricultural raw materials from Sind and the Punjab plains, wood and ivory from the surrounding forests and pearls from the sea.

Trade in these goods—domestic as well as external—became a characteristic feature of the Indus valley people. It was this commerce that sustained the civilization to last so long. However, with the passage of time, no significant change came indicating that a certain equilibrium had already been reached and the basic economic activity continued for centuries without any modification. It is believed that a priestly class managed to extract social surplus in order to maintain a semblance of law and order. The Harappan sites have not yielded any evidence of strong weapons, other than based on bronze, which could have been used to defend it from external or internal threats. The recurrent Indus floods and the inundation of the agricultural fields along the river banks must have washed away the economic base. The threats of external invasions were no less important. The end came abruptly as the Harappan mounds show a sudden break in the layers containing cultural remains. Just as the beginnings of the civilization are obscure, little can be said about the end on the basis of the currently available archaeological data.

Neolithic in the Rest of India : As against this background, the Neolithic cultures in the rest of India, outside the Indus basin, made a late appearance, not earlier than the first half of the first millennium B.C. However, in recent years, more and more evidence has been put forward to suggest that in at least some regions of India the Neolithic cultural tradition started in the pre-Harappan times.

In this connection, one may make a special reference to Burzahom, in the valley of Kashmir, where chronologically the evidence (of period I) goes back to the second half of the third millennium B.C. The tools dug out from the karewa beds at Burzahom include remains of grey and black burnished pottery, stone axes and a variety of stone and bone tools. The evidence also suggests that the Burzahom culture possibly had an affinity with the Neolithic cultures of East Asia, particularly Manchuria. The practice of burying dogs with their masters was a peculiar trait common between the two regions.

The archaeological evidence collected from sites in Southern Deccan (Karnataka), such as Brahmagiri, Sanganakallu, Piklihal, Maski, Tekkalakota, Hallur as well as from the ash mounds of Utnur and Kupgal, along the banks of Krishna and Godavari rivers, reveals a widespread use of ground stone axes. Domestication of animals such as goats, sheep and cattle was common. The Neolithic settlements were nestled on the granite hill slopes which were mostly terraced. The archaeological remains also include handmade pottery, a variety of grey ware. Chronologically, these cultural phases have been placed around c. 2500-1800 B.C.

The cultural succession suggests a second phase (c. 1800-1500 B.C.) in which these earlier cultural traits continue along with the use of copper and bronze. A third phase has also been identified, dating back to c. 1400-1050 B.C. in which the use of metals increases considerably. It has been inferred that these Neolithic communities were mainly dependent on cattle raising, although cultivation of gram and millet was adopted as a supplementary subsistence in the third phase of cultural development. It has also been inferred that in the first phase, the Deccan Neolithic cultures were influenced, particularly in the pottery tradition, by Hissar and certain sites in Baluchistan and Iran. During the second phase the cultural influences seem to have emanated from Malwa, and in the third phase from Maharashtra (*e.g.*, Jorwe type of pottery).

Other Neolithic evidence comes from eastern Rajasthan and Malwa plateau. The archaeological remains collected from Ahar and Gilund in the Banas basin in eastern Rajasthan, suggest a peculiar Banas Neolithic culture with unique characteristics—an absence of stone tools, but numerous copper axes, and a variety of black-and-red ware. It may be noted that copper was locally available.

Nagda and Navdatoli are two sites in the Malwa region which have yielded a variety of artifacts, *viz.*, stone blades, copper tools and painted black-and-red ware, indicating the emergence of the Neolithic culture. These were mainly cattle-raising communities and also domesticated sheep and goats. Chronologically, these cultural phases belong to three different periods which have been assigned dates from c. 2015 to 1380 B.C.

From the geographical spread of Neolithic cultures in the Indian sub-continent, Bridget and Raymond Allchin concluded that Saurashtra, eastern Rajasthan and Malwa appeared to have acted as centres of diffusion of post-Harappan culture in the rest of India. Location ally, these regions were situated at the junction of routes and it is likely that they became scenes of intermingling of different cultural traits—Harappan as well as West Asian. A synthesis of these traditions took place in these nodal regions from where they spread to other parts of India. This diffusion of cultural traits possibly took three routes: a southern route passing through Maharashtra and the Deccan, an eastern route passing through the Narmada and Betwa valleys and a northern route leading to the Ganga valley.

It may be assumed that the late Harappan sites of Rupar (Punjab) and Alamgirpur (Ganga-Yamuna doab) played a significant role in the diffusion of Harappan (and post-Harappan) influences in the Ganga valley. Sites in Maharashtra and the Deccan on which extensive archaeological work has been done bear testimony to this diffusion theory. Sites in the Tapi valley, *viz.*, Prakash and Bahai as well as those in the Godavari and Krishna valleys, *viz.*, Jorwe, Nevasa , Daimabad, Chandoli, Sonegaon and Bahurupa have yielded evidence of a stone industry in association with copper and bronze tools and pottery. These cultural phases have been dated as c. 1700-1050 B.C. The evidence collected from Eran, on the Betwa river, and Tripuri, on the Narmada, indicates the possible routes through which the Malwa influences got diffused into the central and eastern regions of India. These Chalcolithic settlements have been assigned to the period around 1500-1280 B.C. Archaeological remains also show a transition towards the Iron Age around 1040 B.C.

Reference has already been made to Rupar and Alamgirpur. The process of expansion of the post-Harappan influences over the Ganga valley may be traced along the routes connecting Alamgirpur with the rest of the region. It appears that these sites faded away soon after the Harappan phase was over. This is evident from a sudden break in cultural continuity from the late Harappan to the Iron Age. The archaeological findings of this intermediary period consist of copper hoards, dug out from the sites in the Ganga-Yamuna doab, other parts of the upper Ganga valley, Chhotanagpur plateau, Orissa highlands, central Vindhyan region and the Deccan. The ochre-coloured pottery generally associated with the copper hoards is supposed to belong to the late Harappan and the post-Harappan phases. Later excavations at Atranjikhera, in Etah district, confirmed these connections. In fact, a comparative study of the archaeological remains collected from Harappa, Rupar, Autha (Punjab), Alamgirpur, Hastinapura, Atranjikhera, Ahicchetra (doab and the upper Ganga valley), Noh (Indo-Gangetic divide region), Saravasti, Kosambi, Rajghat, Prahladpur, Chirand and Sonpur (middle Ganga valley) corroborate these conclusions.

The different pottery types discovered at these sites help in the reconstruction of a peculiar sequence in cultural succession from ochre-coloured ware (c. 1200 B.C.), painted grey ware (PGW, c. 800-600 B.C.) to the northern black polished ware (NBP, c. 200 B.C.). At many of these sites iron tools have been found in intermediary beds around 1000-800 B.C. These phases have led Allchin to the conclusion that these Chalcolithic cultures of the Ganga valley, east of Punjab, may be assigned to an earlier stream of Aryans who reached the eastern regions prior to the arrival of the Rigvedic Aryans. In the geography of *Rigveda*, Yamuna appears to be the eastern limit. Subsequent streams of the Aryan tribes who moved eastwards are believed to have composed *Yajurveda* and *Atharvaveda* as well as the later texts in regions, lying to the east of Yamuna, from Hastinapura to Magadha.

By now the stage had already been set for the rise of states (kingoms, republics). This was a culmination of the process of colonization of the North Indian Plain which had been going on for centuries.

Sodasa Mahajanapadas : These states emerged as monarchies as well as republics, the latter representing a model of tribal democracy. Scholars of Ancient Indian history, such as Romila Thapar, have tried to trace the transition from lineage to state. Thapar argued that the monarchies were a peculiarity of the fertile Ganga valley. On the other hand, the states that emerged on the northern and the southern periphery of the valley were mostly republics. This distinction owed its origin to the ecological differences as expressed in the surplus generating potential of the two types of ecological regions. Understandably, the republics must have been the first states to emerge.

Gradually, they were transformed into monarchies. Those ousted from the monarchies continued to found new republics on the periphery of the Ganga valley. Thus monarchies and republics had an interesting geographic pattern of distribution. Moreover, the republics were generally controlled by a single tribe. However, in several cases, they were confederacies of more than one tribe. The transformation of republics into monarchies also marked a transition from the egalitarian tribal social order to a caste-based hierarchical structure. Gradually different caste groups got associated with different occupations, leading to a further augmentation of the caste-based society. Over time, caste-specific occupational specializations also contributed to the diversification of economy on a progressive basis, with the cities emerging at nodal points as centres of trade and industry.

Angutar Nikaya, the first Buddhist text, mentions the following 16 *mahajanapadas (sodasa mahajanapadas)*: Kuru, Panchala, Kosala, Kasi, Malla, Vaiji (Vrijji), Magadha, Anga, Surasena, Matsaya, Chedi, Vatsa, Avanti, Asmaka, Gandhara and Kamboja (Table).

Sodasa Mahajanapadas

Mahajanapada	*Capital City/Cities*	*Dialect/Language*
Kuru	Indraprastha/Hastinapura	Kuruvi/Khari Boli
Panchala	Ahichchatra/Kampila	
Kosala	Ayodhya/Saket	Kosali/Awadhi
Kasi	Kasi	Kosali/Awadhi
Malla	Pawa, Kusinagar	Bhojpuri
Vajji (Videha)	Mithila	Maithili
Magadha	Rajagriha	Maghai
Anga	Champanagari	Maghai
Vatsa (Vamsa)	Kausambi	
Chedi		Bundeli
Surasena	Mathura	Braj Bhasha
Matsya	Viratnagar	Bagari
Avanti	Ujjaini	Malwi
Asmaka		
Gandhara	Purushapura/Taxila	Gandhari/Pushto
Kamboja		

The later texts such as the *Digha Nikaya* and a number of *Puranas* repeat this information, although each time new *janapadas* are added to the list indicating the on-going process of colonization of new tracts and the emergence of new territories. Sometimes, the *janapadas* are mentioned as twins (*e.g.*, Kuru-Panchala, Kosala-Kasi, Malla-Vajji, Surasena-Matsaya, Chedi-Vatsa).

This process of coming together of two neighbouring *janapadas* may be a political response to the over-arching authority of certain nodes of power which were emerging as empires. By the time of Buddha, Madhyadesa, or the central region of Bharatavarsa, had already emerged as the major ecumene of Aryan settlement outside the Indus basin. In fact, the Sapta-Sindhava region appears to have faded away as none of the 16 great *janapadas* was situated in this region. Madhyadesa, the 'Middle Country of the Epics' , on the other hand, had as many as 13 of the 16 *mahajanapadas*. The region encompassed the entire north-central zone of India, from the Siwaliks and the Terai in the north to the Vindhyas and the Narmada in the south and from the banks of the Saraswati (Ghaggar) in Rajasthan-Haryana in the west to the Rajmahal hills in the east.

These natural boundaries protected it and made it the most secure region, just as its rich resources of land, water and congenial climate added immensely to its agricultural potential.

The mid-Indian uplands, the Chhotanagpur plateau and the Rajmahal hills defined its southern and southeastern boundary. The deltaic tract of Bengal which emerged as the core of the *Vanga janapada* was a much later development. Anga is not mentioned in the *Angutra Nikaya*. In a fertile plain like this early Aryan settlements must have emerged on the river banks, on the land reclaimed from the surrounding forest. The pioneer villages that emerged on the cultivated tracts finally got organized into territorial units, *i.e.*, clan territories, republics and kingdoms. The rivers and the forests, which occupied vast stretches of land, defined the boundaries between the *janapadas*. These were natural boundaries and they delimited the areas of influence of respective territorial entities. Cities emerged on the river banks and became the foci of political authority. Hastinapura, Ahicchhatra, Kausambi, Ayodhya, Saket, Kasi, Shravasti, Vaishali, Rajagriha, Ujjain and Taxila emerged as nodes of power attracting a large population of landed elites, artisans, warriors and administrators. The Kuru *janapada* lay between the Saraswati and the Ganga encompassing the northern part of the Ganga-Yamuna doab. To its east and the southeast lay the Panchala *janapada*, between the Ganga and Saryu. It also extended between the Yamuna and the Ganga in the middle doab. The Kosala *janapada* extended from the Saryu to the Rapti. Kosala also incorporated the tracts between the Gomati-Ganga and the Saryu reaching as far south as Kasi. The Kasi *janapada* was essentially a city-state based on Varanasi. But it also incorporated the tracts of land to the north of the Ganga in the Ganga-Gomati doab. To the east of Kosala there were four *janapadas* in the Bihar plain, two to the north of the river and two to its south.

The Malla and the Vajji *janapadas* were situated in the northern plain. The former lay between the Rapti-Ganga and the Gandak and the latter between the Gandak and Kosi. Vajji, also known as Videha, was a confederacy of tribes. Magdha and Anga *janapadas* were situated in the south Bihar plain. Magdha extended from the Son to the Mokameh spurs. Anga, an eastern *janapada*, extended upto the Rajmahal hills.

To the west of Yamuna the Surasena *janapada* had its focal point in Mathura. It incorporated the Bharatpur-Dholpur-Karauli region. On the south was Matsaya with its core region lying in Alwar district. Further south was the Avanti *janapada*, occupying the upper reaches of the Chambal and its tributaries on the Malwa plateau and focusing on Ujjaini. To the east of Chambal and south of Yamuna was the Vatsa *janapada*. There were three other *janapadas*, lying outside the Madhyadesa. Two of them—Gandhara and Kamboja—were situated in the north-western frontier region. Gandhara occupied the lower Kabul valley. At times it served as a major link between India and Central Asia. The Kamboja *janapada* occupied the valley of Kunar river on the southern slopes of the Hindu Kush. Asmaka was the southern-most among the 16 *mahajanapadas*. It has been identified with the middle Godavari valley. Madhyadesa became the heartland of the Ganga valley civilization. Gradually local dialects developed with a literary tradition, in certain cases, becoming the repository of folk culture, art, religion and philosophy. This resulted in the process of cultural diversification between the *janapadas*. Each *janapada* evolved its own dialect/language and produced its own *Puranas*. These were the beginnings of cultural regionalism that distinguished India so characteristically. These cultural regions of India are living entities and constitute the basis of regional identity. Centuries of rule from powerful *janapadas*, such as Magadha, could not wipe out this cultural regionalism.

S.M. Ali's *Geography of the Puranas* refers to many other *janapadas* in the western and southern regions of India. The evidence based on Pauranic sources unravels the process of territorialization which continued throughout the proto-historic and early historic times.

Mughal Provinces : One observes a discernible continuity from the *janapadas* to the Mughal provinces. While the political structure underwent many changes in the intervening period of two millennia, it appears that the political divisions were successively built on the legacy of cultural regions. The Mughals built their provincial structure on this very edifice. On the authority of the *Ain-i-Akbari* it is possible to recognize the 12 major provinces in which Akbar's vast empire was divided. It may be recapitulated that the Mughal empire under Akbar extended from the Hindu Kush in the west to Bengal in the east and from Kashmir in the north to the river Tapi in the south. Later, as a result of consistent Mughal effort to extend the imperial frontiers towards the south, three more provinces were added to this list.

The major provinces included Kabul, Lahore, Multan, Dehli, Agra, Awadh, Allahabad, Bihar, Bengal, Malwa, Ajmer and Ahmadabad. The later additions were Berar, Khandesh (renamed as Dandesh) and Ahmadnagar. Each province *(subah)* was further sub-divided into *sarkars* and each *sarkar* into *mahals* or *parganas?* While the *mahals* or *paraganahs* were revenue sub-divisions created for the collection of tax on land, a single *mahal*, or a cluster of *mahals*, having homogeneous soil and climatic conditions was designated as a *dastur?* A separate revenue code *(dastur-al-amal)* applied to each *dastur*. In a way, the *dasturs* may be seen as analogous to agricultural regions. The regional structure of the Mughal empire may be reconstructed by tracing the extent and boundaries of the 12 major provinces. The *subah* of Kabul extended over what is now northern and eastern Afghanistan as well as the North-Western Frontier Province of Pakistan. It also included the territories of Jammu and Kashmir, including the occupied parts of Kashmir and excluding Ladakh. The *subah* of Lahor extended from the river Satluj to the Indus. It incorporated most of the province of Punjab, now in Pakistan, excluding the southern and the western parts.

These southern and the western parts of Punjab constituted the *subah* of Multan. After the addition of Thattah (Sindh) *subah* Multan extended from Ferozpur to the Indo-Afghan borders on the west and from the borders of Rajasthan (the territories of Jaisalmer and Bikaner) to Kachchh and Mekran coast on the south. The *subah* of Dehli lay to its east, extending from Satluj to the Yamuna near Palwal in Haryana and from Rewari to the Kumaon region of Uttar Pradesh. Virtually Dehli incorporated almost the whole of what is now Haryana, National Capital Territory of Delhi, upper doab, Rohilkhand and the hilly districts of Uttar Pradesh. The *subah* of Agrah extended from Ghatampur, on the Ganga, west of Kanpur, to Palwal on the Yamuna and from Kanauj in western Uttar Pradesh to Shivpuri in Madhya Pradesh. The *subah* of Malwah, lying to the south of *subah* of Agra, extended from Kali Sindh to the Aravallis and from Chanderi to the borders of Gujarat near Nandurbar. To the east of *subah* of Agra lay the *subah* of Allahabad. Extending from Jaunpur, north of the Ganga, and from Kanpur, in the lower doab, to the borders of Bihar, Allahabad incorporated the districts of the lower doab as well as the adjoining parts of Bundelkhand both in Uttar Pradesh and Madhya Pradesh. To its north lay the *subah* of Awadh. Extending over the northern and eastern parts of the state of Uttar Pradesh, it stretched from Rohilkhand which was a part of the *subah* of Dehli to the borders of Bihar. The foothills of the Himalayas formed its northern boundary, while the southern boundary was conterminous with the *subah* of Allahabad.

The *subah* of Bihar, broadly speaking, corresponded with the present state of Bihar excluding the Chhotanagpur plateau districts. To the east of Bihar lay the *subah* of Bengal. It extended from Sylhet and Chittagong hills on the east to the borders of Bihar in Garhi on the west and from the sea in the south to the foothills of the Himalayas in the north. The Mughal *subah* of Bengal thus incorporated the whole of what is now Bangladesh as well as the Indian state of West Bengal. The province also included lowlands of Orissa incorporating the *sarkars* of Jalesar, Bhadrak, Katak (Cuttack), Kaling Dandpat and Raj Mahendrih. *Ain* describes the land of the Tipperah tribes (now Tripura) as an adjoining territory lying to the east of the *subah* of Bengal.

The *subah* of Malwah comprised the western and the southern parts of the state of Madhya Pradesh. However, its eastern boundary was vaguely defined. On the east lay the Gond territory over which the Mughals had little authority. The *subah* extended from Chanderi to Nandurbar. The *subah* of Ahmadabad (Gujarat) extended from Burhanpur to Dwarka in Kathiawar and from Jalor to the port of Daman. Its extent and boundaries as described in the *Ain-i-Akbari* correspond admirably with the present state of Gujarat. The *subah* of Ajmer virtually included the dependencies of Amber, Jodhpur, Bikaner, Jaisalmer and Sirohi. Its extreme limits were defined by the *sarkar* of Ajmer in the northeast to Banswara in the southwest and from Pushkar in the east to the western boundary of Jaisalmer on the west. Mewar, comprising the *sarkars* of Chittor and Ranthambhor, shown in the *Ain* as part of the *subah* of Ajmere was, in fact, under the full control of Rajputs.

Mughal Provinces and Sarkars

Subahs	*Sarkars*
Kabul	Kashmir, Pakli, Sawad (Swat), Daur-Banu-Isakhel, Qandahar, Kabul
Multan	Multan, Bhakkar, Thattah, Hajkan, Sewistan, Nasirpur, Chakarhalah
Lahor	Bet Jalandhar Doab, Bari Doab, Rechnau Doab, Chenhat (Jech) Doab, Sindh Sagar Doab

Contd...

Subahs	*Sarkars*
Dehli	Dehli, Badaon, Kumaon, Sambhal, Saharanpur, Rewari, Hisar, Firozah, Sirhind
Ajmer	Ajmere, Chittor, Ranthambhor, Jodhpur, Sirohi, Nagor, Bikaner
Ahmadabad	Ahmadabad, Pattan, Nandod, Baroda, Broach, Charnpaner, Surat, Godhra, Sorath (Kathiawad)
Malwa	Ujjain, Raisin, Garha, Chanderi, Sarangpur, Bijagarh, Mando, Handiah, Nandarbar, Mandesor, Gagron, Kotri Parawa
Agra	Agrah, Kalpi, Kannauj, Kol (Koil), Gwalior, Irij, Bayanwan, Narwar, Mandrael, Alwar, Tijarah, Narnol, Sahar
Illahabad	Illahabas (Allahabad), Ghazipur, Banaras, Jaunpur, Manikpur, Chanadah (Chanar), Bhathkhora, Kalinjar, Kurrah, Karah
Awadh	Awadh, Gorakhpur, Bahraich, Khairabad, Luckhnow
Bihar	Bihar, Monghyr, Champaran, Hajipur, Saran, Tirhut, Rohtas
Bengal	Udarnbar (Tanda), Jannatabad (Lakhnauti), Fatehabad, Mahmudabad, Khalifatabad, Bakla, Purniyah, Tajpur, Ghoraghat, Pinjarah, Barbakabad, Bazuha, Sunargaon, Sylhet, Chittagong, Sharifabad, Sulaimanabad, Satgaon, Mandaran, Jalesar, Bhadrak, Katak (Cuttack), Kaling Dandpat, Raj Mahendrih

It may be interesting to examine the nature of correspondence between the *janapadas* and the Mughal provinces of the sixteenth century. While a broad concordance is there, it is understandable that the *janapadas* expanded in space from their core areas to the periphery during the period of two millennia. The process of peopling continued with the result that the human occupance of land expanded. The extent and boundaries of the core regions also extended. In some regions, such as the the Sapta-Sindahav, where there were no political states as such, new territories emerged.

Mughal Provinces and Sodasa Mahajanapadas

Subah	*Capital City/Cities*	*Corresponding Janapada*
Kabul	Kabul	Gandhara-Kamboja
Multan	Multan	
Lahore	Lahore	
Dehli	Dehli	Kuru-Panchala
Ajmer	Ajmer	Matsya
Ahmadabad	Ahmadabad	
Malwah	Mandu	Avanti
Agrah	Agrah	Surasena-Panchala-Chedi
Illahabad	Illahabad	Vatsa-Chedi
Awadh	Faizabad/Lucknow	Kosala
Bihar	Patna/Azimabad	Malla-Vajji-Magadha-Anga
Bengal	Gaur	Anga
Khandesh	Burhanpur	
Berar	Daulatabad	Asmaka
Ahmednagar	Ahmednagar	

*Anga was not among the *sodasa mahajanapadas* as menhoned in the *Angutra Nikaya*. The colonization of Bengal was a later development.

It may be noted that the Mughal *subah* of Dehli extended over the Kuru *janapada*. Agrah occupied the *janapadas* of Surasena and Panchala. Illahabad focused on Vatsa and Awadh on Kosala. The *subah* of Bihar was an amalgam of Malla, Vajji, Magadha and Anga *janapadas*. The *subah* of Bengal represented a later extension of territory and occupied the Vanga *janapada*. However, Vanga was not mentioned in the *Angutar Nikaya*. The *subah* of Malwah occupied the Avanti *janapada*. In the northwest the *subah* of Kabul had grown out of the core *janapadas* of Gandhara and Kamboja. The Mughal *subah* of Kabul was much larger and marked an outgrowth from the core region of Gandhara. In the process of expansion it embraced the territories of Jammu and Kashmir.

British Ascendancy Influence

The British ascendency in India which became evident with the transfer of *diwani* rights of Bengal and Bihar to the East India Company in the eighteenth century transformed the politico-administrative set-up of the country. Unlike the previous situation the British expansion started from the coastal areas thus giving the littoral region an importance unprecedented in Indian history. Initially, British power was based in the port centres, such as Bombay (now Mumbai), Madras (now Chennai) and Calcutta (Kolkata). Of the three, Calcutta emerged like a colossus replacing the significance of traditional centres of power, such as Murshidabad, Agrah or Dehli. With the location of capital at Calcutta started the process of concentration of political, administrative and financial institutions in and around the new port city. With this the heartland of Hindustan from Lahore to Patna and from Agra to Ujjain lost its former glory.

It emerged as an hinterland of the port centre. The impact of this process of marginalization and under development has not yet been weakened even after half a century of independence. The British also created administrative divisions which had little relationship with the historically evolving ethno-lingual regions. They also imposed a dual administrative set-up of directly ruled provinces and the Indian princely states. Some of their provincial structures were very large. They were created in gross neglect of the ground realities. The British Madras presidency, for example, extended from southern Orissa to include coastal Andhra, parts of Tamil Nadu, Kerala and coastal Karnataka. Likewise, the Bombay presidency extended from Kathiawar and Gujarat to encompass parts of Coastal and interior Maharashtra and Karnataka (Table).

British Provinces and Indian Princely States, c. 1930

British Provinces	*Indian Princely States/Agencies*
Ajmer-Merwara	Rajputana Agency
Andaman and Nicobar Islands	
Assam	Assam
Baluchistan	Baluchistan States
Bengal Presidency	Bengal States
Bihar and Orissa	Bihar and Orissa States
Bombay Presidency	Bombay States
Burman	
Central Provinces and Berar	Central Provinces States
	Central India Agency
Coorg	
Dehli	
Madras Presidency	Madras States Agency
North-West Frontier Province	North-West Frontier Province Agencies
Punjab	Punjab States
	Punjab States Agency
United Provinces of Agra and Oudh	United Provinces State
	Gwalior State
	Hyderabad State
	Jammu and Kashmir State
	Mysore State
	Sikkim State
	Western India States Agency
	Baroda State

The national government in India after independence was confronted with the colossal task of reorganization of states on the ethnolingual principle. However, the task of reorganization of states is still incomplete. The ethnolingual principle may be taken to its logical conclusion to satisfy the aspirations of minority language groups.

2

Geography and Humanbeings

Migration is almost as characteristic of Homosapiens as tool making and culture building. Man is the most widely distributed social animal on the earth surface. From their probable origin in Africa, human groups had spread out to occupy all the major land areas of the earth excepting Antarctica about 20,000 years before present century, long before the beginnings of agriculture and written history.

Thus, migration is a geographical phenomenon that seems to be a human necessity in every age. Since man has a tendency to leave the areas in which life is difficult, he migrates to the areas where life may be easy and better.

Understanding the Concept

Migration has been defined differently by different experts. In its most general sense, migration is ordinarily defined as the permanent or semi-permanent change of residence of an individual or group of people over a significant distance.

Migration may be permanent or semi-permanent. Recently, geographers have been concerning themselves with population dynamics and the problems associated with migration.

Migration together with 'fertility' and 'mortality' is a fundamental element determining population growth and population structure in an area. It is, however, difficult to provide a scientific essential criterion for classification of migration. Migration, in fact, may be international, inter-regional, inter-urban, rural-urban or intra-urban. On the basis of time criterion, migration may be temporary or permanent. If we take into consideration the distance, the migration may be long or short. On the basis of number, migration may be individual or mass; it may be politically sponsored or voluntary. On the basis of social organization, migration may be that of family, clan or individual. On the basis of causes, migration may be economic, social, political or religious. Migration may be stepwise, i.e., village to urban hierarchy.

Causes of Transfer

The causes of migration may be numerous and these may range from natural calamities, climatic change, epidemics, drought to social, economic, cultural and political. The over-population and heavy pressure on resources may be the cause of permanent or temporary, and long distance or short distance migration. Many a time the differences between groups in levels of technology and economic opportunities also cause large-scale migration.

Technology : People with more sophisticated technology may invade and conquer new areas. Contrary to this, less advanced groups may be attracted to the greater opportunities provided by a more developed society. For example, ancient Romans conquered vast areas in Europe, North Africa and South-west Asia. During this period, many people migrated to Rome which provided better economic and employment opportunities. During the medieval period, the Arabs were quite advance in education and technology. With this, they conquered large territories in Central Asia, Northern Africa, Iberian Peninsula (Spain and Portugal) and eastern parts of Europe. In the 14th and 15th centuries, the Europeans had better navigation ships and they discovered America, Australia and numerous unknown islands of the Atlantic, Indian and Pacific Oceans. They colonized and exploited more populated territories of Asia and Africa.

After the Industrial Revolution in 1779, the Europeans emigrated to North America, Australia, New Zealand, South Africa and South America. The large-scale emigration from European countries continued up to the first part of the 20th century and the main reason was to colonize the under-developed countries and to exploit their resources.

Economic Causes : One of the prime motives of emigration seems to be economic. Man's need to have virgin land to till has inspired him to migrate to distant areas. It was because of this reason that the slaves (African Negroes) were transported to the plantations in tropical America. These Negroes subsequently got settled in the United States of America, Latin America and the West Indies. Lust for virgin land also motivated the Europeans to emigrate from U.K., Ireland, France, Germany, Italy, Spain, Portugal, Holland, Belgium and Denmark and to settle into the Praries of U.S.A. and Canada. In the 17th and 18th centuries, about 20 to 40 hectares of land was used to be given free of cost to the emigrants in U.S.A. who owned nothing in their homeland. The temptation to have land became a great magnetic force for the Europeans to settle in America.

Heavy pressure on the land resources in the motherland also forced the people to outmigrate and to settle in areas where economic benefits may be achieved. The pastoral people and nomads of Central Asia invaded the territories of the sedentary people. The Mongols, Tatars and Kurds migrated in the medieval period and got settled in the fertile valleys of Farghana, Panjsheer (Afghanistan), Volga, Armenia and Caucasus mountains.

Non-availability of proper jobs and unemployment are also the economic reasons which compel the youths to leave their home for the places, areas, regions and countries where employment may be found. It was the main cause of the migration of the Scotch-Irish in the 17th and 18th centuries after the decline of the woollen industry and the slump in linen in consequence of the rise of cotton textile industry in Britain. The invention of spinning machine and the establishment of large-scale cotton factories made many of the weavers of the subcontinent of India (India, Pakistan and Bangladesh) unemployed. In 1930, the slump and economic depression resulted into an exodus of British workers to Canada and U.S.A. in spite of efforts to stop them.

The agricultural labourers, if unemployed, also leave their native places. In every age, the labourers emigrated to neighbouring or distant fertile tracts. For example, in about 2000 B.C., Abraham migrated from Ur (Mesopotamia) to Cannan, the Greeks migrated to Scythia, the Jews migrated to Nile Valley and Mesopotamia, and the Bihari, Orissan and Rajasthani labourers migrated to West Indies and South-east Asia. At present, labourers from Bihar, Uttar Pradesh and Rajasthan are migrating to Haryana, Punjab and Himachal Pradesh.

Thus, the economic poverty, unemployment and attraction for better economic opportunities always motivated and forced the peoples to emigrate from their native places. For example, poverty of agricultural labourers in West Bengal, Orissa and Bihar resulted into the emigration waves towards the metropolitan cities and the productive plains and agriculturally developed parts of the country. The regions and districts of marginal farmers and small farmers also emigrate with the agricultural labourers as the tiny size of their holdings is unable to provide them adequate sustenance.

Overpopulation : An excess of population in an area in relation to resources and available technology is known as over-population. Over-population may exist at local, regional and national levels. At present, it is most frequently seen in under-developed rural areas where the outstripping of resources by population growth may be evident in under-nourishment or under-employment.

Throughout the human history, migration took place because of the overpopulation in a community or region. In such a case, emigration may affect all social classes. At present, overpopulation is the most important cause of emigration in the developing countries of Asia, Africa and Latin America. The people of these countries are emigrating to Saudi Arabia, Kuwait, U.A.E., Canada, Greece, France, Germany, Australia and New Zealand. The emigrants include domestic servants, agricultural labourers, unskilled workers, technicians, engineers, doctors and academicians. Each year, from India alone, about 3 lakh skilled and semi-skilled people outmigrate.

Social and Religious Causes : The human desire to stay, work and enjoy life with the people of his ethnic, social and religious groups is also an important cause of migration. In every period of human history social factors led to large-scale emigration. In the middle ages, there was the emigration of Balkan peoples owing to the dominance of Muslim Turks. In the 20th century, there has been expulsion of Jews from Germany, Spain and Russia (erstwhile U.S.S.R.). There is large-scale outmigration of Muslims from Bosnia and Serbia (erstwhile Yugoslavia). Muslims are moving out from Myanmar (Burma) to Bangladesh. The feeling of insecurity is compelling many of the Kashmiri Pandits and Punjabi Hindus to outmigrate from Jammu & Kashmir and Punjab respectively, while the

Muslims prefer to migrate from the Hindu dominated areas to Muslim localities and vice versa, irrespective of the social amenities. Many of the well-off Muslims are moving towards the Muslim slum localities in Delhi, Mumbai, Ahmadabad, Meerut, Allahabad, Agra, Kanpur, Lucknow, etc., because of social and religious persecution and same is the case of Hindus as they are outmigrating from the Muslim dominated areas to Hindu dominated areas. Health, climate, education and other social amenities are also responsible for migration at the regional, national and international levels.

Political Causes : One of the important causes of migration, especially after the Second World War, is the political one. Political refugees is a worldwide phenomenon today. One can list numerous examples of political migration. These included, Turkish, Armenians, and white Russians early in the 20th century; European Jews after the Second World War, Palestinians, Chinese, Hungarian (freedom fighters), Cubans, Indians, Pakistanis, Bangladeshis, Kashmiris, Tamils, Vietnamese, Afghans, Iranians, Somalians, Kurds, Serbians, Bosnians, Kosovos and Albanians. All these are the examples of forced migrations induced by political factors.

Another form of involuntary migration is the expulsion or exchange of minorities by nations. For example, the Sudeten Germans repatriated from Czechoslovakia after the Second World War, and the Muslims and Hindus exchanged when India and Pakistan were created by the partition of the sub-continent in 1947. The forced and compulsory migration is always bound up with tragic highlights in human history. During the medieval period, Negroes used to be sold in the urban markets of Portugal, Spain, France and Italy. These slaves (Negroes) were exported to U.S.A. from the western coast of Africa, from the Senegal to the Gold Coast and Congo. The settling of the New World (after 1492) opened up new markets, but it was Spain which needed them in the Americas.

Demographic Causes : A number of demographic factors also play a vital role in the migration pattern. For example, age has been recognized as one of the important demographic factors controlling the degree of desire to move among the potential migrants. It is not surprising that adults are more migratory' than other age groups. It is the rate of growth of population that determines the extent of population pressure in an area. The great historic movement of the Europeans across the Atlantic Ocean was an expression of increasing pressure of population on the resource base of Europe. Similarly, in India, the large-scale denigration from the densely populated parts of Orissa, West Bengal, Kerala, Bihar and Uttar Pradesh is largely due to a poor population-resource ratio in these areas.

Diffusion of Information : The availability of information through education, cultural contacts and spatial interaction also increase the chances of population migration. The communities that are ruled by orthodoxy, conservativism, traditions, customs and strong communal ties are less mobile than those which are socially more awakened, progressive and have more contacts and exposure with the outside world. The information network and cultural contact increase the horizons of job opportunities. Thus, migration generates more migration, which signifies the role of information network in the stimulation of migration. In India, the Sikhs are the most adventurous and well-informed people, who migrate even to the less developed and less attractive areas like Bolivia, Columbia, Ecuador, Honduras, Nicaragua, Central African Republic, Chad, Sudan, Ethiopia, Yemen, etc.

General Rise in the Level of Aspiration : With the advancement in science and technology, new items of comfort and luxury appear in the market and the level of aspiration of the educated and uneducated people goes up. Everybody is tempted to enjoy a better standard of living. In India and Pakistan as well as in all the developing countries, the young men who were better-off than their fathers were nonetheless dissatisfied, and many sought to better themselves overseas. It is mainly because of this reasons that Indian engineers and doctors are emigrating to U.S.A., Canada and in large number unskilled and semi-skilled labour in migrating towards Saudi Arabia and the Gulf countries.

Wars : Wars have been one of the important causes of human migration. Wars have always involved upheavals, particularly in the regions where these have been fought. The First World War (1914-1919 A.D.) involved the displacement of about 6 million people, and the Second World War (1939-1945 A.D.) involved the involuntary displacement of about 60 million people. Some people were forced to move to avoid political and religious persecution even long before the war. Millions were moved in the forced transference of ethnic minorities, millions more in evacuation and flight from the battle fronts. Forced labour movements and deportation accounted for several millions, and subsequent resettlement involved still further movement. Nearly a million Poles and Jews were deported by Germany during the war, another million by Russia, who also moved a large number of Germans from the Volga to the far eastern parts of Siberia. The population of Europe was in constant flux during the war, from the Volga to Paris, and the end of the war led to the movement of refugees. All these migrations resulted into the changes of human resources, in land use, customs, traditions and human values.

After the Turko-Greek war of 1921, about 0.35 million Turks were moved to Turkey from Greece and about 1.2 million Greeks went to their own country from Turkey. As stated above, transfer on a much greater scale followed the partition of India, where approximately six million Hindus left Pakistan and an equal number of Muslims left India for Pakistan though large minorities remain in both countries. In the summer of 1999, millions of Muslims from Kosovo were forced to migrate to Albania and Mecedonia. After the attack of U.S.A. on Afghanistan in October 2001, millions of refugees migrated to Iran, Pakistan, Tajakhistan and Uzbekistan.

Government Policy : The government policy of a particular country also favourably or adversely affected the pattern of population migration. The British, French, German, Russian, American, New Zealand, Canadian and South African governments have specific population policies and most of them discourage immigrants. In the early fifties, a large number of people moved out from the countryside to the urban areas as a result of collectivization of farms. Similarly, the political totalitarianism in Russia resulted in the migration of people from their homeland at the time of Bolshevik Revolution, also known as the Great October Revolution (1917).

The Results

By the process the migration or the outgoing population influences the society, economy and environment both at the places of origin and destination. These effects may be favourable or harmful to the ecology, society and economy. It has been rightly pointed out by some of the leading experts of population that each migrant by nature seeks to create something of the original milieu in the midst of the new environment and consequently enriches the culture and civilization. The main consequences of migration have been briefly described in this section.

The place of origin, the place of destination, the individuals and families of the migrants undergo a qualitative and quantitative change in their population and demographic structures. In general, people from the crowded and over-populated regions outmigrate to the areas of scanty population with strong resource base. In other words, it leads to reallocation of human resources with a view to achieve a better balance between human resources and physical resources.

Consequently, the population-resource relationship of the two areas involved in the process of migration gets modified significantly as a result of the movement.

The migration of people from one place to another brings a tangible transformation in the demographic characteristics. The absolute number of population, its density, growth, fertility, mortality, age, sex, literacy and occupational structure get transformed. At the place of immigration, an increase in the density of population brings more pressure on resources. The new arrivals may also enhance the capacity to exploit its resource potential more scientifically. The denigration of the educated and technically trained people into an area improve the literacy rate and quality of life in the region.

In both the regions (place of origin and place of destination), the emigrants effect the society, demography, culture and economy. This point may be explained by citing an example. Suppose if 'X' persons leave country 'A' and migrate to country 'B', what change takes place in the size of the two populations. The commonsense answer that the population of country 'A' will decrease and that of country 'B' will increase. This generalization is, however, true only in short run. If the typically young migrants have their children in their new country, its fertility rate may go up, while that of their native country goes down. Since the remaining population of country 'A' will then be older on the average, its death rate may go up, while that of country 'B' goes down. In short, after one generation, the transfer of 'X' persons will in fact amount to 'X' plus a certain proportion based on the migration's effects on the population structure, and rates of population growth of the two (A and B) countries. Thus, the population structure, sex composition and work force of both the countries, regions will be substantially affected.

While the resources, their utilization, society and economy are affected, the emigrants also face a lot of problems of adaptation. The people who migrate from the rural areas to the urban industrial areas suffer from the lack of open space, good housing and fresh air. The highly polluted atmosphere of the new place of work (urban centres), the toxic fumes and dust affect their health adversely. They also have to adapt to the new dietry habits and timings of food. There are evidences to prove that the incidence of respiratory diseases, cardiac problems and cancer among the people who migrate from rural areas to industrial towns is very high. Sometimes, the physical contact of the people belonging to two different racial or ethnic backgrounds may change the biological characteristics of the people. Greater interaction between the people of Africa (Negroes) and that of Americans and Europeans have affected the racial characteristics of these people. Consequently, some of the diseases like AIDS, which were inoffensive formerly, have become now deadly.

When large-scale immigration takes place as it happened during the 18th and 19th centuries in America in which millions of people emigrated from European countries to Anglo-America and Latin America, the cultural values of the two groups of people underwent substantial transformation. The immigrants bring with them their language, religion and cultural values. Integration and amalgamation of people become serious problems when the people having different languages stay together. For example, the French and English in Canada; the Dutch and English in South Africa; the Indians, Pakistanis and Bangladeshis in South-west Asian Arab countries; and the Tamils and Sinhalese in Sri Lanka are confronted with such type of problems. As a matter of fact, the migrants not only try to preserve their religion, culture, language and customs but also try to spread them. The people belonging to different religious groups, whenever move and start living with the people professing other religious faiths, it may either mean a healthy spread of all religions in areas or may cause people of one religious faith to flee for fear of suppression and persecution by the others. The immigration of the Catholic, French and British Protestants in Canada and the multitude of Asian religions to U.K. are the examples where a healthy spread of different religious faiths has emerged. Contrary to this, the Jews and Arabs in West Asia provide an example of the latter category where followers of one religion have been suspicious of their persecution and suppression.

The dietary habits of people at both the ends are also substantially influenced. For example, the Indians and Chinese who migrated to U.K., U.S.A. and other countries opened the restaurants and hotels in which they serve the dishes of their countries. It is because of this reason that Idli, Dosa, Biryani, Tandoori, Kabab, Samosa, Rasgullah and Barfi are increasingly becoming popular among the British and Europeans. Similarly, the Chinese dishes like Chowmein has gained popularity in the western world. The continental breakfast (bread, butter, toast and eggs) is becoming more popular in urban India, Bangladesh and Pakistan which provides an example of the influence of English rulers on the upper and middle classes of their colonies. The interaction of people of different ethnic and cultural backgrounds when stay together lead to the enrichment of civilization. This may be said as the greatest benefit to human society from migration.

Global Shifting

On the basis of volume (number), distance and means of transport use, international migration may be classified into: (i) inter-continental, and (ii) transoceanic. The international migration is quite older, and limited in volume and time. It is caused because of a number of special conditions at the place of origin and place of destination. International migration may be short migration of small groups of people or large groups travelling long distances. Contrary to this, the transoceanic migration has been observed largely in modern times, affecting the chief centres of population in Europe, America, India, China, Japan, South-east Asia and South-west Asia.

Reliable data and information on international migration are not availably. In ancient time, there had been large-scale exodus of people to Canaan where Jews immigrated. The Bactria and Sogdiana were the promised land for the Persians where they immigrated in large numbers. The Mongols migrated to South China and Thailand around 2000 B.C. and the people from Central Asia arrived in waves in the sub-continent of India. In India, millions of people from peninsular India migrated to the tea plantations in Assam and Bengal. In the beginning of the 20th century, about 30 million Chinese invaded Manchuria and settled there. Similarly, people from China outmigrated and settled in the different countries. From South-west Asia, people migrated towards Sudan and the Muslim Fulani moved westwards to the oases of Chad, Algeria, Nigeria, Niger, Mali, Central African Republic, Mauritania, Congo basin and Cameroon. Along the southern coast of the Mediterranean Sea, the Arabs penetrated towards Libya, Algeria, Morocco, Spain, Portugal and France. In America, both before and since the arrival of the Europeans, there had been much migration among the Americans. Until recently, the Amerindian (Red Indian) tribes often migrated in the United States, Canada, Mexico, Central and South America.

There is much internal migration still in India, Indonesia, and Americas. In West Indies and Brazil, the seasonal labourers go to the centres of banana production. From Mexico there is a constant stream of workers to the United States. The Negro workers migrate from southern states to the industrial centres in the north-east and the Great Lakes regions. Similarly, in India, agricultural labourers from Bihar and eastern Uttar Pradesh seasonally migrate to Punjab, Haryana, Jammu & Kashmir and western Uttar Pradesh. There had been waves of migration in Europe. In fact, European emigration is the most tremendous displacement of people ever recorded by history, especially after the 17th century. The numbers involved have never been counted. It began well before the 19th century with the discovery of America and its colonization by Spaniards; but it has been estimated that between 1821 and 1910, at least 26 million migrants left Europe for the United States alone. According to one estimate, since the end of the 15th century at least 105 million persons left Europe for the Americas, Asia, Africa and Australia. In the 16th century, 3 million Spaniards left home for America. Upto the 19th century, only the Spaniards, Portuguese, English, Scots, Irish, Greeks, German, Italian, Belgian and French nationals were outmigrating. The destination of all these peoples was mainly the new countries. Upto 1914 there was a considerable flow to the United States. In 1921, the immigrants quotas were fixed by an Act. The Europeans outmigrated to many countries in Asia, Africa and Australia.

The European emigration occurred mainly because of over-population and industrialization. The Europeans emigrated in two different directions:

(i) They entered into the sparsely populated tropical and subtropical coast-lands which were easily accessible and which possessed potential for the production of exotic crops because of their hot and humid climates. Consequently, commercial production of sugar, cotton, tobacco, coffee, tea, spices, indigo, rice, and so on flourished in the eastern coastlands of North and South Americas. In order to cultivate these virgin lands labour had to be imported. Initially this labour was supplied by the Europeans themselves but as the demand started multiplying, it gave rise to slave trade from Africa. After the slavery was abolished, the densely populated countries of Asia were exploited by the British and Dutch colonialists. Consequently, the semi-slave trade supplied labour to newly developed plantation agricultural areas of Malaya, Sumatra, Sri Lanka, Fiji, Hawaii, East African countries and Mauritius.

(ii) Secondly, large European emigrants moved into the temperate zone of the Americas, South Africa, Australia and New Zealand.

Although people emigrated from England, Scotland, France, Holland, Germany, Belgium, Spain, Portugal, Italy and other European countries, the denigration from Ireland was almost one of the unparallels in the human history. The great exodus was due to an extraordinary increase in the population of Ireland. The country was over-populated, the physiological density was very high and the people were dependent largely on potato cultivation. The failure of potato crops for successive three year (1845, 1846, 1848) and higher prices for their meagre food meant famine for the whole country. This was the start of rightful poverty and of a vast exodus, a real flood of emigrants, which in some years carried off more than a third of the population. Between 1850 and 1900, more than 40 lakh people left Ireland. The check on the movement which began in 1885 was due to depopulation of the country, to a dizzy fall in the birth rate.

Emigration from the British Isles as a whole represents the greatest swarm of mankind that has ever crossed the sea. About 17 million persons left the island between 1815 and 1926. This steady flood of emigrants has not diminished the population of Britain where an excess of births over deaths has constantly kept up the total and even increased it.

The first half of the 20th century may be said as a period of great political turmoil in which the two world wars were fought. The First World War resulted into forced labour. The forced migration because of political crisis and wars resulted into great human tragedies. The significant characteristic of the forced labour is that the normal selectivity in migration by age, sex, skill and education is lacking and communities as a whole are uprooted. Over 20 million Russians migrated within and outside Russia after the Great Revolution of 1917. In 1947, partition of the sub-continent of India resulted into the migration of about over 12 million people from one part of the sub-continent to another.

At present, the number of international migration has declined because of strict immigration policies. The migration is controlled and regulated by the policies of the governments and the international regulation. There are very few countries in the world which allow or encourage the immigration of foreigners on a permanent basis. The number of those countries which still attract a sizeable immigrants is small, which include U.S.A., Canada, Australia, New Zealand, Greece, Germany, Saudi Arabia, U.A.E., Central Africa and Israel.

During the last three decades, the South-west Asia has emerged as an important region which is attracting technical, skilled and unskilled labourers. Among these countries, Bahrain, Iran, Iraq, Jordan, Kuwait, Libya, Oman, Qatar, Saudi Arabia, U.A.E. and Yemen are important. There are refugees and illegal migrants also who cross the international boundaries. The emergence of Bangladesh and Bosnia as well as the Gulf War of 1991 resulted into enormous movement of refugees across the international borders.

Population Redistribution through International Migration : During the period between 1821 and 1932 (between the Nepollonic Wars and the Great Depression) there was relatively unrestricted, but reasonably well-recorded, mass movement. At least 59 million people, mostly Europeans, moved overseas during the period. Out of these, over 90 per cent settled either in North or South America. During the 20th century, the largest volume of population transfers had been through forced migration negotiated and organized by national governments. After the First World War over a million Germans moved into Germany from Alsace-Lorraine and the areas ceded to Poland. In the exchange of population between Greece and Turkey decreed by the 1923 Treaty of Lausanne, about 1,200,000 Greeks were moved out of Asia Minor and Eastern Thrace, and about 600,000 Turks were moved from various Balkan countries, mostly from Greece. The German conduct of the Second World War resulted in vast movements of refugees and forced labour, and the collapse of Germany set in motion a gigantic series of redistribution of population. The partition of the sub-continent into India and Pakistan in 1947 led to an exchange of population estimated in 1949 at over 6 million in each country.

The first phase of emigration from Asia to countries of America, Australia, and Eastern Africa began as a result of the abolition of slavery in the British colonies in the 1830s. The demand for plantation labour was met by the recruitment of indentured labourers from India, and later from China, Japan and Philippines. Although this class of socially induced migration dominated Asian migration throughout the 19th century, it particularly ceased after the 1920s.

The decline in international migration has been offset by an increase m interregional migrations within Asia. Immigrants have contributed greatly to the economic development of Malaysia (Chinese and Indians), Myanmar (Burma) (Indians), Sri Lanka (Indians), Borneo (Chinese) and Indonesia (Chinese). International movements of migrants within Asia continued throughout the 1920s and 1930s, despite the near cessation of mass migration in the West. The chief countries of emigration have been China, India, Japan, South Korea and the Philippines. For all the migrations, principally to Eastern Africa and South-east Asia, the effects on the population of the country of origin have been negligible. For example, the urban areas of Japan between 1920 and 1940 increased by 17.5 million persons, ten times the number emigrating Japanese civilians. As another example, about 30 million Indians are estimated to have settled abroad between 1834 and 1937, but since 24 million eventually returned to India, the net outward balance was only 6 million over the span of more than a century.

Since 1945, with the gaining of political independence by former colonies, the new Asian nations have focused attention on internal migration and the achievement of economic development by labour of their own people rather than by that of immigrants. As a consequence, international migration in Asia has sharply declined.

The period immediately following the Second World War saw the revival of mass international migration in the West, including the transfer of some 1,200,000 refugees. International migration during the period from 1945 to 1952 may be summarized in the tabular form as follows:

International Migration

Continent	*Number*
Emigration from Europe	4,452,000
Immigration into Europe	1,150,000
Other intercontinental migration	710,000
Total	6,312,000

Source : Spencer et al. (1954).

The chief countries of emigration and the main receiving countries during 1945-1952 have been given below in Table:

Migration from Europe (1945-1952)

Emigration	*Immigration*		*Number*
Great Britain	1,107,000	U.S.A.	1,104,000 (27%)
Italy	741,000	Argentina, Brazil and Venezuela	883,000 (21%)
Netherlands	318,000		
Spain	272,000	Canada	726,000 (17%)
Portugal	152,000	Australia	697,000 (17%)
		Israel	526,000 (13%)
		South Africa	125,000 (3%)
		New Zealand	75,000 (2%)

Source : Spencer et al. (1954).

It may be seen from Table that North and South Americas continued to receive more than two-third of the 'new' inter-continental migrants.

Demands in Vogue

International migration, which includes both voluntary migration for economic or other reasons as well as the involuntary movement of refugees, is on the rise. Data are uncertain and trends are difficult to track, but, according to the United Nations, at least 120 million people (excluding refugees) lived or worked out side of their own country in 1990, an increase from about 75 million in 1965. The annual growth rate of immigration has been steepest in the developing countries, and about half of all international migration takes place within the developing countries.

So far as the involuntary migration (refugees) in concerned, it was about 18 million in 1993 which has fallen to 14 million in 1996. However, the number of refugees is overshadowed by the increase in the number of internally displaced persons—those who have been forced to flee their homes by armed conflict, persecution, or natural or man made disaster, but who remain within their national borders. Because of the rising number of civil wars and local conflicts, the number of internally displaced persons now totals an estimated 30 million worldwide, mostly concentrated in 35 countries. Africa is the worst affected region, with upto 16 million people having been internally displaced. Several million Muslims migrated from Kosovo (Yugoslavia) to Albania and Mecedonia in 1999, because of persecution. About 4 million Afghans took shelter in Pakistan, Iran, Tajakhistan and Uzbekistan during the period between 1996 and 2001.

Environmental degradation and resource scarcity can help to trigger mass migration. Population growth, land scarcity, and cycle of droughts and floods has encouraged the illegal immigration of more than 10 million Bengalis—perhaps 20 million including their descendants—to neighbouring Indian states of Assam, Tripura and West Bengal. The influx has prompted local resentments, and more than 4,000 people were killed in a series of violent clashes in the early 1980s. Tension continues today.

Many governments increasingly view immigration as a problem, despite the fact that immigrant labour often benefits both the home and host countries. Perception of national identity, cultural differences, and fears of unemployment are all contributing to actual and potential hostilities between immigrants and nationals. Nonetheless, immigration, either legal or illegal, seems certain to continue.

Leaving the Country

India has the second largest population in the world. The Sutlej-Ganga plains, the coastal regions of the peninsular India, and the states of Kerala, Maharashtra and Gujarat are particularly the areas of very high density of population. In fact, in many areas, India is overpopulated and the population exceeds the capacity of the land to feed it. People, even in the less crowded areas like Rajasthan, Himachal Pradesh, Laddakh and Arunachal Pradesh, are poor. In the historical past also, the major concentration of population had been in the Great Plains of India. Undernourishment in India is traditional and becomes intolerable during the years of drought and famine. Yet, the cultivators are attached to their villages often by religious or social ties. As a result of this attachment with Dharti Mata (Mother Earth), emigration was considered to be the worst misfortune, especially in the Hindi belt.

Emigration, especially movement of the Indian people at substantial scale in India, began after the arrival of British. The British started emigrating the Indian agricultural labourers into the sugarcane growing areas of the tropics in which it was difficult for the Europeans to work in the fie'ds under the hot and humid climates. In general, the Indians are hardworking and acclimatized to work in the hot and moist climates. They are also eager to make money, live on the minimum and can work even at low wages as they are poor and many a times find no employment in their own towns and villages. They are thus ready to emigrate even to the distant places. In the 19th century, the English people started emigrating to those colonies in which sugarcane, tea, spices, banana and rubber were the dominant commercial crops.

The first group of Indian labourers was emigrated in 1815. It consisted of convicts and criminals, transported from Calcutta (now Kolkata) to Mulieratus. Mauritius became a real training ground for agricultural labourers on the sugarfields. After 1934, many free workers were attracted by the mild and oceanic climate of Mauritius. Today, out of the total population of 1.2 million persons, there are more than 0.7 million Indians in Mauritius. Hindi and Urdu are followed and spoken by most of the people of Indian origin. For the last several years, even the prime minister of this country is Indian by origin. A big colony of about 8,000 Indians exists in Reunion Islands situated near Mauritius in the Indian Ocean. The Reunion Islands also have extensive cultivation of sugarcane. People of Indian origin got settled in the rubber producing areas of Malaysia and sugarcane growing areas of West Indies.

In the continent of Africa, the Indians emigrated since the Arabs active trade in the Indian Ocean. With the help of the Arab traders, the Indian merchants established their business in the countries of Africa. The Europeans also encouraged the Indian labourers to work in their fields, to establish plantations, to clear forests for agricultural development, and to construct railways. The labourers were followed in large number by the other social elements, such as traders, small shopkeepers, moneylenders and usurers. As a result of these emigrations, there developed large Indian colonies in Natal, Kenya, Uganda, Tanzania, Somalia, Mozambique, Zimbabwe, Yemen, U.A.E. and South Africa. Most of these Indian emigrants are from the states of Gujarat, Maharashtra and Punjab.

After the Second World War, many of the colonies of Europeans were decolonized and independence was achieved by most of the African countries. Owing to these political developments, many of the Indian immigrants were expelled from the African countries. In fact, the Indians were treated as foreigners, and most of the African countries are virtually closed to Indian immigrants.

Indians began to reach the West Indies around 1840. Their number was the largest in British Guinea and Trinidad, where there are over two lakh Indians in each. They have also settled in Jamaica, Martinique and Guadeloupe in small numbers, but homogenous groups. In Fiji, the Indians first arrived in 1874. At present, they number nearly three lakh and constitute more than half the population. These emigrants are mainly from Bihar and Uttar Pradesh.

At first the Indian emigrants were the labourers engaged by the professional recruiting agents, but the Emigration Act of 1922 enabled emigration to be controlled by the Government of India. Subsequently, at the invitation of relatives or friends, emigrants freely and voluntarily followed in the track of former indentured labour, and little Indian communities sprang up everywhere with their tradesmen, craftsmen, and in some places their priests of temples and mosques.

By the end of the 19th century, the emigration movement became quite numerous. The tea and rubber estates of Malaya (Malaysia) and Sri Lanka (Ceylon) also attracted the Indian workers in large number. In Sri Lanka, there are more than 3.8 million Indians which constitute about one-third of its total population of about 16 million (1991). There are about 14 lakh Indians in Malaysia out of its total population of 21.5 million

(1998). Most of the immigrants in Sri Lanka and Malaysia are from the state of Tamil Nadu. Besides, there are numerous Indians who settled in Myanmar (Burma) and Nepal. These people outmigrated from the densely populated parts of Bihar and eastern Uttar Pradesh.

Concepts of Shifting

As stated at the outset, the Homosapiens is a culture building migratory species. Right from his appearance on the earth surface, throughout the Paleolithic and Mesolithic periods, man migrated to kill beasts; in the Neolithic period he migrated to search for lands which may be brought under cultivation; during the historic periods he sailed to have trade relations with the neighbouring countries and to conquer the new lands; in the medieval period he migrated in search of new resources, to establish his colonies and to avoid persecution; while in the modern and contemporary periods man is migrating either to improve his economic status or to stay at a place where he feels socially and culturally more secure and politically more independent. The causes of migration as discussed at the outset are numerous and so are the consequences. The scholars of population have done a lot of empirical research and tried to establish certain laws of migration. Some of the important laws of migration have been given in the following paras.

The first scholar to formulate laws of migration was E.G. Ravenstein who based his generalizations on empirical studies of population movement in Britain, the United States and some of the countries of Northwest Europe. He observed the migration data between 1885 and 1889 and arrived at certain very important conclusions which may be taken as the laws of migration. Subsequently, other leading scholars like Lee corroborated his findings and added some new laws of migration. The laws of migration, advocated by Ravenstein, are as under:

***The Majority of Migrants go only a Short Distance (distance decay law)* :** This law seems to have been operational at least since the medieval times and is still operative today. With the advent of modern transportation, the average distance travelled by migrants may have increased, but relatively short moves are still the most common. For example, the first major concentrations of the American black (Negroes) population in northern cities were in Baltimore, Washington and Philadelphia, all of which are relatively close to the source areas of migrants who left the south in large numbers after 1910. There is a tendency that most of the migrants move short distances whereas few go long distances. This tendency is known as distance decay. This expression applies to most of human activities. In fact, there is a decline in the amount of some phenomena with increasing distance from a focal place. In other words, many human activities and features tend to cluster near accessible places, so their frequency, volume or value usually declines with distance from the point of attraction. This tendency applies to the land values around a market, population densities ranging an urban centre, number of migrants to a place of attraction, and numerous other phenomena that are affected by spatial interaction. As an example, consider the number of migrants to city X who came from districts A-G. When the amounts of migration are plotted on a vertical axis and the corresponding distances are displayed on the horizontal axis, the graphed trend indicates that migration declines (i.e., 'decays') as distance increases.

***Migration Proceeds Step by Step* :** Ravenstein's second law of migration is that the inhabitants of the country (rural area) immediately surrounding a town of rapid growth flock into it; the gaps thus left in the rural population are filled up by migrants from more remote districts (rural areas), until the attractive (gravitation) force of one of our rapidly growing cities makes its influence felt, step by step, to the most remote corner of the region.

Distance Travelled by Migrants to City X

District Name	*Distance from City X (kilometres)*	*Number of Migrants (per year)*
A	15	1000
B	30	500
C	45	350
D	60	250
E	75	200
F	90	175
G	105	150

Accordingly, sequential moves extend the effects of migration spatially. Such a series of moves was an important feature of migration to the American frontier in the 19th century when farmers who wanted to move on to new lands would often sell out to later migrants. On a different scale, a series of step by step residential shifts is usually generated in urban areas today when a family moves into a newly built house and thus vacates an older house, which is then reoccupied by another family (which leaves another residential gap and so on).

Migrants going Long Distances Generally go by Preference to One of the Great Centres of Commerce or Industry : This tendency of moving towards the great centres of commerce and industry has been operative since medieval times, when London attracted population from all parts of England. The pull of large cities is apparent in the developing countries. The industrial and commercial centres of Tokyo, Mumbai, Kolkata, Delhi, Karanchi, Lagos, Sao Paulo, New York, Manila, Jakarta, Seoul, Shanghai, Mexico, Rio de Jeneiro, Djakarta, Bangkok, Tehran, Cairo, Dhaka and Nairobi, Hyderabad, Bangalore, Ahmedabad, Kanpur and Pune.

The effects of a large place on the size of the migration field is expressed by a Gravity Model. This mathematical model states that the number of migrants to a place is directly related to the population size of that place but inversely related to the migratory distance.

Each Current of Migration Produces a Counter-current of Lesser Strength : This law of migration seems to be universal and applicable in almost all the developed and developing countries. Even in the extreme case of slave trade (15th-19th centuries) produced a tiny counter flow back to Africa of people who, in one way or another, were able to regain their freedom and to return to their homes.

Migrants who choose to move long distances to new places often move back. For example, of the 13 million migrants to the United States from 1900 to 1914, an estimated four million returned to Europe during the same period. In recent years, the proportion of emigrants from the United States has increased relative to the number of immigrants into the country, with the largest numbers returning to Mexico, Germany, Canada and the United Kingdom.

The Natives of Towns are less Migratory than those of Rural Areas : This observation was related to a stage of economic development in Europe in the 19th century when rural-to-urban migration was predominant. In most of the developed and developing countries today, movement is still mainly from rural to urban places.

At present time, most of the developed countries have large urban majorities and relatively small rural populations. Therefore, most migration is interurban or intra-urban (within a city). In these developed countries, there is an increasing trend of outflow from urban centres to rural areas. This migration is associated with the decentralization of industrial jobs and the willingness of commuters (daily passengers) to travel long distances to their urban places of work.

Females Migrate more Frequently than Males within the Country of Birth, but Males Frequently Venture Beyond : This law of migration was related partly to a stage of economic development and partly to its particular cultural context. Since women had few employment opportunities in rural areas, they tended to migrate to cities.

At present, the gender of the majority of migrants depends on cultural conditions as well as on economic and employment opportunities. In India, for example, many more males migrate from rural areas to cities than do females. During the British period, Calcutta (now Kolkata) and Bombay (now Mumbai) were the main centres of commerce, trade and industries and attracted the people from distant places of Bihar and Uttar Pradesh. Nowadays, all the million cities of India, particularly Mumbai, Delhi, Kolkata, Chennai, Hyderabad, Ahmedabad, etc., are attracting male migrants from all corners of the country. Migration between rural Indian villages, however, is commonly made by females because brides traditionally move at the time of marriage to the village of bridegrooms.

Man is more active in international migration than woman. In the Gulf countries and Saudi Arabia, there are large number of immigrants from India, Pakistan, Bangladesh, Sri Lanka, Indonesia, Philippines and South Korea and over 85 per cent of them are male workers.

Most Migrants are Adults; Families Rarely Migrate out of their Country of Birth: This law contains two observations. The observation concerning adults is universal and indisputable. In voluntary migrations, the majority of people are adults. The second part of the law is more problematic. It is certainly true that families find it more difficult to move than unmarried adults, but owing to the cultural, religious and political factors, families migrated from one country to another country. Many refugees migrated from India to Pakistan and Pakistan to India in 1947, and from Bangladesh to India in 1971. Similarly, families migrated from Ireland to Canada; from Uganda and Kenya to U.K. and U.S.A.; from Somalia and Ethiopia to Yemen, Egypt and Saudi Arabia; and from Palestine (west bank of Jordan and Gaza Strip) to Syria, Lebnon, Libya, Kuwait and Saudi Arabia.

Large Towns Grow more by Migration than by Natural Increase : It is universally accepted that large cities grow more fast because of inmigrants and population influx. For example, over 60 per cent of the total population of Delhi, Mumbai and Kolkata belongs to people who came in these cities from the distant parts of the country in search of employment and got settled.

Most of the large towns and cities in the developing countries today are growing very rapidly by the inmigration of people from the rural areas. In addition to job opportunities (which have always attracted migrants to large cities), today's major cities of the developing countries offer more medical/ educational facilities and social entertainment.

The Main Causes of Migration are Economic : This law is also universally accepted. It has been argued that international migrants tend to be influenced more by conditions in the area of destination than by the pressure of population at home. The movement of Indian labourers to Saudi Arabia and the Gulf countries is mainly because of unemployment in rural India and better job opportunities in the South-west Asian countries.

It is, however, difficult to verify that the major reason for migration is economic. Human motivation, decisions and behaviour are so complex that it is very difficult to compare the relative importance of economic factors with other considerations. For example, moving at the time of marriage is not regarded as primarily an economic move in those societies having love marriages. Also many persons with wealth or large pensions enjoy the flexibility of choosing their residence in areas having an attractive physical and/or social environment.

Most of the above discussed laws of migration, advocated by Ravenstein, are universally accepted. There are, however, many questions which Ravenstein did not address. For example, the non-economic, cultural, social, political, psychological and religious causes of migration have not been examined by him. Since the time of Ravenstein several other theories and generalizations have been propounded by other scholars to explain the process of migration.

Migration Model of Lee : The famous sociologist Everett Lee propounded another theory of migration in 1965. He generalized following four factors which influence the decision of an emigrant:

— factors operating in the area of origin,

— factors operating at the destination,

— factors that act as intervening obstacles, and

— personal factors that are specific to individual.

The potential migrant is influenced by positive and negative factors associated with both the place of origin and the possible destination. Lee suggested that the potential migrant weighs the known and expected advantages and disadvantages of the destination in comparison with the situation at the place of origin. For a move to result, the attractions of the destination must be great enough to outweigh the advantages of staying and to over-come any intervening obstacles, such as distance, cost of relocation, and the disruption of established patterns of life.

Personal factors also affect the evaluation process. For example, a family with children in school, college, or a couple without children, may not consider the quality of public schools, colleges and universities at either the point of origin or destination, but these educational institutions may be an important factor in the balance sheet of a family with children who are in the education stage.

Similarly, attachment to place and proximity to family and friends may prevent a potential migrant from moving, even when other factors, such as promotion opportunities seem attractive.

Mobility Transition Model of Zelinsky : In 1971, Zelinsky, the famous expert of population and migration also propounded a theory about migration. His theory of migration is known as the Mobility Transition Model.

Zelinsky proposed that changes in migration behaviour have been paralleled by the stages of the Demographic Transition Model. Similarity between the two models is not surprising because demographic conditions and migratory decisions are both related to changes involved with the urbanization, industrialization and modernization process.

The trends in volume of migration through the five stages of the demographic transition model at four different scales:

(i) international,

(ii) regional,

(iii) rural to urban, and

(iv) urban to urban including intra-urban.

In addition, Zelinsky prepared graphs to show the interrelationships between migration and circulation (short term generally repetitive movements) over the five stages.

The mobility transition, as with the demographic transition model, is primarily a depiction of stages experienced by modernized societies. Nevertheless it may provide a basis for anticipating some future migration patterns in countries undergoing economic and demographic change.

In the first stage of mobility transition, when population growth was negligible because of high death rates cancelled out high birth rates, little migration occurred at any scale. In India, in 1931, for example, only 10 per cent of the population lived outside the district of their birth. Life was localized, information on other places was in short supply and most people lived and died in the locality of their birth. Some population movement did take place, but relative to later stages of the demographic transition model, migration was less common. Circulation was mostly daily trips to fields and occasional journeys to markets and festivals.

The second stage when population increased rapidly because the death rate dropped while the birth rate remained high, was a time of great migration. Mounting population pressure on the land, better transportation system, and a widening sphere of exploration and trade, and bringing knowledge of other places gave rise in the past to increased population movements at all scales. People migrated from one country to another, from settled areas to new frontiers, from rural areas to growing cities, and from towns to cities. Emigration from Europe and 19th century domestic migration to American frontier illustrate these conditions.

Zelinsky's third stage is transitional, matching the third stage of the Demographic Transition Model when the birth rate began to fall toward the death rate and population growth rates declined. International migration lessened and agricultural frontiers closed, as in the United States by 1930. But, at the same time, rural to urban migrations and movements within and between cities became more important. With the shift to secondary and tertiary occupations, people increased their circulation by commuting to jobs and travelling to obtain special services (e.g., medical and educational) and so on.

By the fourth and fifth stages, when low birth and death rates brought little population growth to the new developed society, migration was predominantly interurban and intra-urban. Some international movements occurred, particularly of skilled and unskilled workers from the less to the more developed countries, but progressively stricter immigration laws limited the number of people who could cross international boundaries legally. The rural to urban movement virtually ceased because the percentage of farming population, and hence potential movers was very low.

As would be expected, circulation increased as the societies modernized. The development of the private automobile greatly widened the radius of daily travel. When circulation increased, migration became less imperative. For example, by commuting an hour each way every day, there is an area exceeding 20,000 square kilometre in which to find a job. This increased access to jobs reduces the need to migrate. It may be summarized that as the economic inter-dependence increases and standard of living rises, more and more people outmigrate from the places of their births and countries.

3

Geography and Population

Global Scenario

The spatial distribution of population is one of the most important topics of human geography. It is significantly important, because man has brought tremendous change over the earth surface. At present, man is the single most important powerful geographical factor transforming the earth surface at an unparallel pace. Scientific and technological achievements of man have reached dizzy heights and nothing now seems to be beyond his reach. Man has walked on the surface of the moon. Scientists are at the threshold of achieving superconductivity, which will usher in revolutionary changes in global civilization. Man has tools to change the climatic conditions. Man's scientific advancements have resulted into acid rain, global warming, atmospheric pollution, ozone depletion, nuclear winter and nuclear radiation and has further threatened to destroy nature's ecosystem across the world. The study of human population and its spatial patterns are thus of vital importance.

In the initial phase of human history and prehistory, the human population grew at a snail speed. The hazardous climatic conditions, the migratory character of hunters and gatherers, and the poor nourishment were all unfavourable for the growth of population. During the last ten thousand years, the growth of world population accelerated at two distinct times. The first time was at about 10,000 years B.P. (before present) or 8000 B.C. when man started the domestication of plants and animals. This abrupt change in human activity is also known as Agricultural Revolution. The assured supplies of food from agriculture and settled settlement life provided better nourishment and the human body became more resistant to the adverse weather and climate. Consequently, the human population started increasing at a steady pace. The second time when the population increased faster was in 1779 when man harnessed fossil fuels and developed steam engine. This development is known as Industrial Revolution.

Both the above mentioned developments in the human history had profound effects on the spatial distribution of population. These revolutions altered the population patterns and demographic attributes of the world as well.

The agricultural and industrial revolutions are important ecologically because population numbers and densities are at the heart of relationships between humans and natural environment. If population numbers rise or fall, everything else is liable to change, including land use, settlement forms, economy, migration patterns and social mobility. Before discussing the present distribution and density of population, it is of great interest to have a brief description of human population prior to the beginning of agriculture.

Population Prior to Agricultural Revolution : There is no direct or documentary evidence to know the growth, pattern and density of population of the world during the prehistoric period. The scholars of population have, however, estimated the population of the past on the basis of circumstantial evidence. For example, agriculture was unknown before about 8000 B.C.; prior to that all human groups made their living by hunting and gathering. Not more than 52 million square kilometres of the earth's total land area of about 150 million square kilometres could have been successfully utilized in this way by our early ancestors. From this and the population densities of hunting and gathering tribes of today, it has been estimated that the total population in 8000 B.C. was about 5 million people (Durand, 1973). They lived in bands of fewer than fifty people; the men being primarily hunters, whereas the women providing most of the food through plant gathering. Nowhere on the earth was the population large. Typical densities were four persons per 100 square kilometres. There were large uninhabited areas, and each band usually lived in relative isolation from others. Such low densities are not surprising because the economies of bands were based on a form of extensive land use. In other words, they obtained their food from large areas without expanding much labour per unit of land.

It was the period of high birth and death rates. The high death rates restricted population growth. It is probable that population growth was also restrained by deliberate control through restricted sexual intercourse, abortion, infanticide, and the use of plants with contraceptive powers. Slow growth for a band meant that its size did not rapidly exceed the carrying capacity of the territory being used for hunting and gathering (Deevey, 1960).

The distribution of population prior to the development of agriculture covered only parts of the old continents. By 8000 B.C., hunters and gatherers had migrated from Africa throughout Europe and Asia, to Australia, and across the Bering Straits and southward the length of the America. Only Antarctica was totally uninhabited by mankind.

Growth of World Population

Year	*Population in million*	*No. of years to add one billion people*
10,000 B.C.	5	
1A.D.	200	
1000 A.D.	300	
1750	800	
1850	1,000(1 billion)	80
1930	2,000(2 billion)	32
1962	3,000(3 billion)	13
1975	4,000(4 billion)	12
1987	5,000(5 billion)	10
1999	6,000(6 billion)	13
2012	7,000(7 billion)*	14
2025	8,000(8 billion)*	25
2050	About 10,000(10 billion)*	

*Printed figures.

Source : United Nations Population Reference Bureau, 1998.

The size of population at various times from the onset of Agricultural Revolution (8000 B.C.) until the first scanty census data (recorded in the 17th century A.D.) has also been estimated. It is thought that the total human population at the time of Christ was around 200 to 300 million, and that it has increased to 500 million by 1650 A.D.

In 1850, the world's population was estimated to be 1,000 million. It means it was doubled from 500 million in 1650 to 1,000 million in 1850. The population doubled again to 2,000 million by 1930. In 1975, it rose to 4,000 million mark and in 1987 became 5,000 million. In 1999, the world population crossed the mark of 6,000 million, and around 2012 it will be about 7,000 million (Table). The projected population for 2025 and

2050 is about 8 billion and about 10 billion respectively. The pattern of human population growth. It may be seen from this that the size of the population has increased continuously and that the rate of increase has also risen stupendously.

According to the estimates, the population, on the average, doubled once after every 1,500 years during the Neolithic Period (New Stone Age). The next doubling, from 500 million to 1 billion, took two hundred years and the doubling from 1 billion to 2 billion took only eighty years. The population reached four billion around 1975, having doubled again in only forty-five years. If the present growth rate of population continues, it will become 7 billion in 2012 to reach 8 billion in 2025 and about 10 billion in 2050.

Doubling Time of World Population

Period	*Population*	*Time in which population doubles*
10,000 B.C.	5 million	
1650 A.D.	500 million	1500 years
1850 A.D.	1000 million	200 years
1930 A.D.	2000 million	80 years
1975 A.D.	4000 million	45 years
2025 A.D.	8000 million*	43 years

* Projected figure.

Source : United Nations Population Reference Bureau, 1998.

Papulation : Historical View

As described in the preceding paragraphs, the population of our ancestors a few million years ago was confined to Africa and numbered only in lakhs. By the time our ancestors invented agriculture, the information started passing from generation to generation. The transmission of knowledge about hunting, gathering and preparation of food and identification of enemies helped in the expansion of agriculture and growth of population.

The development of agriculture around 8000 B.C. resulted into the decline of death rate. It is guessed that the increased food supply led to better nutrition, greater resistance to disease, increased longevity of life and hence a growth in population. This may be correct, but we have no conclusive proof that death rates fell. They may have even risen shortly after the advent of agriculture. Disease spread more rapidly in closely packed communities than in small and relatively isolated hunting and gathering bands.

It is also possible that increase in birth rates after the advent of agriculture resulted in some social changes. Rules and sanctions that had been used to keep hunting and gathering bands small and in balance with their resources may have been relaxed in agrarian societies that required more labour to produce food.

There are other evidences also which prove that the population after the advent of agriculture increased at a faster pace. Using evidence from living populations of hunters and gatherers such as the San and Kung (Bushmen of Kalahari desert), supporters of this view have concluded that our ancestral hunters and gatherers consciously tried to space the births of their children. Child-spacing among nomadic hunters and gatherers was presumably necessitated partly by the inability of the mother to carry more than one child.

Another likely reason would have been the need to nurse each child for at least three years because the environment lacked the 'soft-food' required for earlier weaning. Moreover, it appears likely that prolonged breast-feeding itself may help in suppressing fertility. Infanticide may also have been practised widely among hunter-gatherers, when children were born too soon and in time of scarcity. The Polynesians, who live in the islands in Pacific Ocean, also practised infanticide, polyandry, non-marriage of landless younger sons, abstention from sexual activity for a period of time after birth of a child, coitus interruption and abortion. It seems from these evidences that hunter-gatherers practised child-spacing.

When man started agriculture, the imperatives for child-spacing disappeared and population started growing at a faster pace. It is generally agreed that agriculture and high natality go hand in hand. The prime reason for this is the perceived economic value of children to farming families. In other words, in agrarian societies, children are commonly viewed as economic bonuses. They serve as extra hands on the farm and take care of parents in their old age. The growth of human population was, however, not continuous after the Agricultural Revolution.

Civilizations rose, flourished and disintegrated; periods of good and bad weather occurred; and famine and war took their toll. Despite fluctuations in the birth and death rates, agriculture permitted the existence not only of higher population densities and settled village life but also of large-scale cooperative ventures, specialization of labour, development of crafts (pottery making etc.) and social stratification. The growth and development of irrigation and the emergence of towns and cities concentrated economic power in the hands of numerically small elite, who controlled much of the land and food resources. The great mass of society, however, consisted of peasants and labourers who worked the land and produced the food surpluses upon which the entire social structure was built.

During the medieval period of human history, there was more emphasis on commerce and trade. This resulted into growth and development of towns and cities and the consequent increase in the demand of consumer goods. It was the period when the landowners wished to put more land to the plough.

Gradually, in Europe, agriculture started taking the shape of big business. The life expectancy (the number of years an infant in expected to live), however, remained below thirty-five years. The doubling time for the world population was about 1,000 years.

The main characteristics of population of the world prior to the Industrial Revolution were high infant mortality, short life expectancy, periodic famines and undernourishment, vulnerability to epidemics, high birth and death rates, and wide fluctuations of death and responding birth rates. The outcome of these conditions was that, although there were sharp periods of population decline, they were followed by short-term revival. The result was a slow, long-term increase in population numbers.

The second major turning point in the population growth came with the Industrial Revolution. The Industrial Revolution led the systematization of production with the help of power-driven machine. This resulted into high output per capita and rapid accumulation of wealth. The distinguishing feature of the Industrial Revolution was that fossil fuels, primarily coal, provided an energy source other than animal muscles, wind or water power. The use of fossil fuels allowed productivity to increase rapidly because human effort was supplemented by powered mechanical devices.

The Industrial Revolution in Europe and North America created a new society in which the primary activities (agriculture, forestry, mining, fisheries) began to decline and the secondary (manufacturing) and tertiary (services) activities began to take an increasingly important role. This led to diversity in national economy. Machines were increasingly utilized to supplement or replace human effort. These machines required the application of scientific knowledge to design and maintain them. Family became as the unit of production. Goods were produced for sale in regional, national and international markets. There occurred specialization in economic activities. The population was tempted to migrate from rural to urban areas and the productivity per capita increased substantially.

At the time of the Industrial Revolution, urban population increased rapidly, and living standards in new industrial towns were abysmal, especially for the poorer families. Consequently, disease flourished in over-crowded, unsanitary urban settings, and death rates were often catastrophic when cholera contaminated water supplies. Similarly, the manufacturing processes produced wastes that polluted the air, land and water. In the early phase of Industrial Revolution, the education system was not well-developed, literacy rate was low, medical facilities were not adequate, and the per capita income and the standard of living were low. The Industrial Revolution resulted in an uneducated peasantry migrating into the expanding urban centres created numerous socio-economic and cultural problems.

Thus, it is clear that the world's population has greatly increased in the last few centuries, but the rate of increase quickened only after 1900 and has continued to quicken during the last four decades, particularly after the Second World War.

The tremendous growth of population may be appreciated from the fact that in 1950 the total population of the world was more than 2,523 million which rose to more than double in 48 years, being 6 billion in 1999. The estimated population of the world at the present growth rate would be about 8,039 million in 2025 and about 9,366 million in 2050.

It may also be observed from Table that the population growth rate would be highest in Asia, followed by Africa and Latin America. There would be a marginal increase in the population of Anglo-America (U.S.A., Canada) and Oceania, while the population of Europe will decline from 729.40 million in 1998 to about 701.07 million in 2025 and 637.58 million in 2050. The negative growth of population in Europe is a point of great concern for the Europeans. Most of the European countries have, consequently, adopted the policy to give incentives to the people to increase population.

World: Size and Growth of Population

(in million)

World/Continent	*1950*	*1998*	*2025**	*2050**
World	2523.87	5929.83	8039.13	9366.72
Asia	1402.02	3588.87	4784.83	5442.56
Europe	547.31	729.40	701.07	637.58
Africa	223.97	778.48	1453.89	2046.40
Latin America	149.29	462.59	641.40	754.20
Anglo-America	171.61	304.07	369.01	384.05
Oceania	126.12	294.60	40.68	45.68

* Projected figures.

Source : World Resources, 1998-99, World Resource Institute, UNO, New York.

Selected Countries: Size, Growth and Douhling Period of Population (2000)

Country	*Population (in million)*	*Years for population to double*
China	1277	61
Irulia	1027.0*	35
U.S.A.	281.4	100
Indonesia	212.1	43
Brazil	170.1	40
Pakistan	156.9	27
Russia Federation	146.9	
Bangladesh	129.2	29
Japan	126.3	270
Nigeria	111.5	26
Mexico	93.0	31
Germany	81.2	
Iran	62.2	25

Contd...

Country	*Population (in million)*	*Years for population to double*
Egypt	60.1	31
U.K.	58.6	280
France	58.0	182
World	6055.0	43

*Assuming current growth does not change projected figures.

** in 2001

Sources : 1. Human Development Report 2000, Oxford University Press, New Delhi, pp. 223-26. 2. Census of India, 2001.

Each year, the world population grows by almost 90 million—an increase of 2.8 people each second. The world population will continue to grow at least until 2050.

It may be observed from Table that China with a population of 1,277 million or 21.50 per cent of the world population (2001) ranks first, followed by India with 1027 million or 16.87 per cent (2001) of the world population. The third and fourth positions are occupied by U.S.A. (281.4 million) and Indonesia (212.1 million). Brazil, Pakistan, Russia and Bangladesh rank fifth to eighth, followed by Japan (126.9 million).

The doubling periods of population of the selected countries between 2001 and 2025 have also been given in Table. It is interesting to note that the population of Iran will be doubled in only 25 years, followed by Nigeria (26 years) and Pakistan (27 years). According to the given estimates, the population of Bangladesh and India would double in 29 and 35 years respectively.

It may also be seen from Table that the populations of U.K., Japan, France and U.S.A. are growing at a snail pace, as the doubling period of population in U.K. is 280 years, followed by Japan (267 years), France (182 years) and U.S.A. (100 years). Russia and Germany are the two countries in which the population may not double even after five hundred years. Contrary to this, if the negative growth rate of population in Russia, Germany and other developed countries continues, the greater parts of Europe would be depopulated. It is in this background that the population policies are being formulated in European countries to encourage the growth of population. While providing incentives for larger families, the European governments may think to liberalize their immigration policies.

Table that contains data on the size, growth and percentage share of populations of some of the selected countries, reveals some interesting features. For example, in 2000, the total population of China was 1,277 million or 21.50 per cent of the world population. In 2025, the population of China will grow to about 1,480 million but its percentage share will come down from 21.50 to 21.14. In 2001, India's population was 1,020 million, constituting 16.87 per cent of the world population, which shall rise to 1,330 million in 2025, accounting to 19.38 per cent of the world population.

Contrary to this, the percentage share of U.S.A. which was 5.0 in 2000 will decline to 4.74 per cent in 2025. The share of Japan and Brazil in the World population will also decline in the coming decades, while Pakistan and Bangladesh will record an alarmingly stupendous growth in population. The Population of Pakistan in 2000 was 156.5 million which is likely to become 2,069 million in 2025 and its share in the world population will go up from 2.6 per cent in 2000 to about 3.93 per cent in 2025. A similar trend is expected in Bangladesh where the population will rise upto 180 million in 2025, having a share of about 3 per cent of the world population.

Selected Countries: Size, Growth and Percentage of Population

Country	*Population (in million)* 2000	*Percentage*	2025*	*Percentage*
China	1277	21.50	1480	21.14
India	1020**	16.87	1330	19.38
U.S.A.	281.4	5.00	332.4	4.74
Indonesia	212.1	3.50	275	4.07
Brazil	170.0	2.80	216	2.80
Pakistan	156.5	2.60	268.9	3.93
Russia	146.0	2.43	131	1.99
Bangladesh	129.2	2.15	180	2.96
Japan	126.9	2.11	121	1.75
World	6000.0	—	8039	2.96

* Projected figures. ** in 2001.

Sources: 1. Human Development Report 2000, Oxford University Press.

2. Census of India, 2001.

Factors Influencing the Distribution of World Population

The unevenness in the distribution of world population may be attributed to the following factors:

(i) Availability of arable land and water.

(ii) Age of civilization.

(iii) Accessibility of places.

(iv) Restrictions of national boundaries.

Arable Land : There is a close relationship between the arable land and the concentration of population. The plain areas having fertile soils and suitable climate (temperature and rainfall) for the cultivation of crops are the regions of high density of population. Still about 60 per cent of the world population is dependent on agriculture for their livelihood. Consequently, the areas of intensive agriculture have high density of population. In other words, there is a positive correlation between the distribution of arable land and the distribution of the world's population. It is not only the plain areas where agriculture is possible, because man through his efforts has developed agriculture on terraces on mountain sides (e.g., Angami tribe near Kohima in Nagaland), heating greenhouses in cold climates (e.g., Netherlands, Germany, France, Denmark, Norway, Sweden, etc.), and irrigating deserts (e.g., Nile Valley, Turkmenistan, Jaisalmer and Bikaner districts of Rajasthan, India etc.).

Nevertheless, a comparison of the distribution of arable areas and population reveals several similarities. In the great plains of India and the eastern plains of China, there are high densities of population. Same is the case with irrigated parts of Nile, Syr and Amu valleys, where irrigation allows farming in what is climatically deserts.

Age of Civilization : The second most important factor which influences the growth, density and concentration of population is the age of civilization. Generally, the longer a place has been continuously used by farmers, the dense and large is the population. The Eastern China plains and the Indo-Gangetic plains have long history of intensive agriculture and they have high densities of population. Contrary to this, the plains of Mississippi in U.S.A., Pampas of Argentina, Down grasslands of Australia and New Zealand and Velds of South Africa though equally productive, there the cultivation was started only after the 17th century. Consequently, these areas are relatively sparsely populated.

Accessibility : Accessibility of places and regions is also a very important determinant of population growth and its distribution. In the agriculturally less productive but industrially advanced societies, the people are dependent on secondary and tertiary sectors and most of the industries have been located at places which are easily accessible. Accessible places are those which are easily connected by transportation to many other places. The factor of accessibility is roost important in an economy based on manufacturing and trading. The economic advantages of locations like Kolkata, Mumbai, Chennai, Karanchi, Rotterdam, Chicago, Rome, Alexandria, Djakarta, Aden, Lagos, Singapore and Tokyo create employment opportunities that attract large populations. These centres of dense population grow both by natural demographic increase and by producing the markets that generate even more economic opportunities for additional population. Consequently, these agglomerations where people have been engaged in industrial activities for several generations (Western Europe) tend to be densely populated.

Restrictions of National Boundaries : Each country has its international boundaries. Crossing of these boundaries by the people of one country to another country is not allowed by the proviso of International Law. Owing to this restriction, the people of overpopulated countries cannot enter the developed countries with less density of population. For example, Bangladesh has a large population density while Australia, Canada and U.S.A. have low density of population. In the absence of restrictions of national boundaries, the population from countries like China, India, Pakistan and Bangladesh could have migrated to U.S.A., Canada and Australia to earn a better living. Such international migration is very limited, mainly because most nations do not allow a mass influx of new immigrants. Countries like Australia, Canada, New Zealand, etc., have enormous resource potential but they do not allow immigrants unless they are technically and educationally outstanding. In other words, political restriction to international migration of people is an important factor which helps to explain the existing distributional pattern of world population. Although major migrations have altered the pattern of the world's population in the past, such movements 'have been drastically curtailed in recent decades. Most governments restrict immigration, and several countries control emigration as well.

According to World Resources 1998-2000 (World Resource Institute, 2000), the population of the world has crossed the mark of 6 billion in 1999, but what is more significant in relation to future food production is the rate of population growth that has characterized the post-1950 and which forms the basis for the exploration of global and inter-regional population growth rate into the 21st century.

It may be seen from table that population growth rate has increased dramatically since 1950, doubling in just thirty-seven years which is in contrast to the previous doubling period of population from 1,250 million to 2,500 million years B.P. The growth rate of population even between 1850 and 1950 was much slower. These trends are startling, and while they reflect increased life expectancy and improved health care, they also reflect increase in pressure that has been brought to bear on earth resources and on food producing system.

It indicates that the most significant population increase has been in the developing nations where some 75 per cent of the world's population is now concentrated. Although global population growth rates are gradually declining and many nations, especially in the developed world, they have undergone a demographic transition from a state of growth to one population equilibrium) there are still many developing countries where population growth rates are in excess of 3 per cent per annum.

Most of these countries are in Africa in which populations are expected to be doubled in about twenty-three years. Elsewhere in the developing world, population growth rates are declining; in South America, the declines have been small, while in China the population growth has been halved in the last decade (1991-2000). The life expectancy is below 40 years in Sierra Leone (38 years), whereas in Guinea and Gambia 40 years each. The average age in India and Pakistan is 69 and 70 years respectively, while it is 79 years in France, 77.2 years in Germany and 80 years in Japan.

Growth Rates of Population of Some Developed and Developing Countries

Country	*Life expectancy at birth*	*Annual growth rate of 2001 (%)*
Bangladesh	58	2.0
Nepal	57	2.1
India	69	1.6
Pakistan	70	2.4
Indonesia	67	1.1
Nigeria	56	2.2
China	70	0.7
South Korea	72	1.0
U.K.	77.2	0.0
Japan	80	0.0
France	79	0.3
Germany	77.2	-0.01
U.S.A.	77	0.7
Russia	67	-0.2

Sources: 1. Human Development Report 2000, Oxford University Press, New Delhi, pp. 223-26.

2. Census of India, 2001.

There are also striking variations in the population composition of the developing and the developed world. According to the estimated population of 2001, in the developed countries, only 20 per cent of the population is in the under 15 age-group and 15 per cent in the over 65 age-group, while in the developing countries 36 per cent of the population is under the age of 15 and about 6 per cent is aged over 60 years.

Thus, improving health care, coupled with high fertility rates and a large proportion of people entering their reproductive years, provide an inbuilt momentum for future high population growth even if fertility rates decline.

There are 19 countries in Europe in which more than 15 per cent of the total population is over 65 years of age and only about 20 per cent is around or below 15 years of age. These are: Austria, Belgium, Bulgaria, Croatia, Denmark, Finland, France, Germany, Greece, Hungary, Italy, Latvia, Norway, Portugal, Spain, Sweden, Switzerland, Ukraine and the United Kingdom.

Solidity of Population

The spatial distribution of world population is not uniform. There are wide regional variations in the density of population. According to 2001 population data, published by the United Nations, the average density of the world population is 41 persons per sq km. Asia, with a density of 108 persons per sq km is the most densely populated continent of the world. Europe got the second rank with a population of 101 persons per sq km. The continent of Latin America, with a density of 21 persons per sq km, has the third rank, which is followed by Africa, Anglo America and Oceania, having a density of 20, 14, and 3 persons per sq km respectively.

Population Scattering

The single most striking fact about the world population is that it is not uni-formly distributed. Moreover, the population distribution has continuously changed in space and time, with migration and varying rates of population growth. In reality, nearly half of the world population is clustered on just 5 per cent of the land, while about 33 per cent of the total land area is virtually uninhabited. The spatial distribution of people with its great unevenness is one of the important questions of human geography that demands exploration discussion and explanation.

If we glance at the examine a map of world population, it becomes clear that there are certain regions which are densely populated and other areas which seem to be empty and underpopulated.

Densely Populated Regions : There are four areas in the world where the average density of population is more than 100 persons per sq km. These are:

1. East Asia (China, Japan, South Korea and Taiwan).
2. South Asia (India, Pakistan, Bangladesh, Sri Lanka, Maldive and Nepal).
3. Northwest Europe (U.K., France, Germany, Netherlands, Poland, Belgium, Luxembourg, Ireland, Denmark, Spain, Portugal and Italy).
4. Eastern North America (North-east United States and South-east Canada).

All these areas of dense population lie in the northern hemisphere and are so placed that more than 75 per cent of the world's population is now concentrated between the Tropic of Cancer and 70° north. Of these, China and the subcontinent of India were having large populations from the earliest times. Europe is less ancient, and the United States have become densely populated only during the last two hundred years.

Except in Japan, the population of East and South Asian countries is living mainly in rural areas. The people are directly or indirectly engaged in agriculture. They are largely dependent on primary activities (agriculture, forestry, fisheries, mining) which reflect the food-producing potential of the land. In these countries also, the areas having fertile alluvial soils, gentle slopes and available water (from surface and underground or rainfall) have the highest densities of population.

In China, the valleys of Yangtezekiang, Hawang Ho and Sikiang have great clustering of population. In India, the Sutlej-Ganga-Brahmaputra plains, the eastern coastal plains and the valleys of the perennial rivers are densely populated. In Pakistan, the province of Punjab, traversed by the five rivers (Indus, Jhelum, Chenab, Ravi, Sutlej) and the delta of Indus are densely populated. The entire alluvial tracts of Bangladesh, excepting parts of the Chittagong Hills, have high density of population. The valley of Nepal and the coastal areas of Sri Lanka also have heavy concentrations of population.

In contrast to the distributional pattern of population is South East Asia, the European and North American areas of high population densities consist of urban agglomerations. The industrial development and tertiary activities (services) in cities and towns have drawn people from the villages and countryside over the last two hundred years, so that over three-fourth of the European and American populations live in cities.

The highest population densities by country occur not in China and India but the highest urban densities are found in Singapore (100%), Kuwait (98%), Belgium (96%), the Netherlands (91%), and the United Kingdom (90%). The European and American population agglomerations could became large because of certain other favourable physical factors. The climate of these regions is mild and temperate in which a variety of crops can be grown. Wheat, barley, oats, rye, suger beet, potatoes, fodder, maize, fruits, grapes and oilseeds grow well up to Ireland, Scotland, Sweden, Norway and Faroes Islands (Denmark). The rise, development and expansion of large-scale manufacturing enabled working people to collect in masses around factories and caused the growth of cities. The cumulative improvement of the various kinds of work, and consequently of the workers, shows that the European population agglomeration is the result of physical and cultural factors. Trade relations and colonization of the African, Asian and American countries greatly contributed to the increase in population of Europe. The expansion of British, Dutch, French, German, Portuguese and Spanish empires in other continents enabled the imports of food and raw materials. The fast growth of European population thus may largely be attributed to industrial development and exploitation of resources of the colonial countries for food and raw materials.

The densely populated regions of North America have a very short history. Up to 1492, America was not known to Europeans and it was Columbus who discovered it (1492) and provided a virgin land for the growing population of Europe. The population of America started growing fast after the Industrial Revolution (1779). The North American concentration developed as an outlet for the European population. By 1750, small numbers of British, Dutch, German, French, Spanish, Portuguese and Italian emigrants had settled along the north-eastern coast of North America. They superseded the native Americans (Red Indians) who were displaced west-ward and were decimated by war, disease and other traumatic effects of cultural contact. The urbanized belt from Boston to Baltimore in the United States and the areas along the St. Lawrance river in Canada have remained the main regions of attraction for the European people. In fact, this belt is the hub of economic, commercial, cultural and political activities in North America.

Apart from the above described four major regions of dense population, there are isolated and scattered nuclei of dense population. Among these small centres of dense populations, the deltas of Mekong, Menam, and Irrawaddy rivers as well as the Indonesian island of Java are noteworthy. Africa's most densely populated areas include the linear concentration along the Nile Valley in Egypt, the ring of settlement around Lake Victoria, and the coastal areas of Nigeria. The Latin American areas of high population concentration include the central part of Mexico, where nearly half the nations population lives. In Central Mexico, rural densities exceeds 2,600 persons per square kilometre, and Mexico City (with a population of over 20 million) is one of the leading world metropolis: The coastal areas of Vene-zuela, Brazil, and Argentina, with cores around the cities of Caracas, Rio-de-Janeiro, Sao Paulo, and Buenos Aires, respectively, are also areas of high population density.

Sparsely Populated Regions : As stated at the outset, over 70 per cent of the land surface is sparsely popuated. In general, the hot, cold, arid and mountainous areas are sparsely populated, which may be classified under the following categories: (i) Deserts and arid lands. (ii) Ice caps and cold regions. (iii) Mountainous regions. (iv) Tropical rain forests.

Deserts and Arid Lands : All those areas where the rate of evaporation is higher than the ra:e of precipitation (arid lands) are most sparsely populated. The main problem with the settling of arid lands is the deficiency and non-availability of water. In the arid areas where water is available, agriculture is often successful because desert soils are generally rich in minerals and in many places the growing season is long. The Nile Valley, the Indus Valley, the Sonara desert in Mexico and United States and other scattered oases in Central Asia are the main exceptions to the general condition of barrenness and relative emptiness. Excepting these fertile valleys and oases, the great series of deserts, especially along the Tropic of Cancer and Tropic of Capricorn on the western margins of continents, the population is most sparse. In the deserts of Libya, Algeria, Egypt, Niger, Mali, Sudan, Chad and Mauritania, the average density of population is about one person per 15 sq kms. In such deserts of hostile climate and scarce water, people deliberately check the number of births by imposing monogamy or even celibacy on large scale proportions of their numbers, as is done in Laddakh and Tibet. In some cases, this anxiety of over-population may lead to the rise and survival of barbarous customs. For example, in some Somali tribes, until very recently, a man might not get married until he had killed with his own hands a certain number of men of his tribe.

In brief, the deserts (both hot and cold) are almost uninhabited or else thinly populated by nomadic herders, hunters and gatherers (Bushmen of Kalahari), and nomads (Badawins of Arabia and Sahara). In Central Asia and the peripheries of Australian desert, the communities keep sheep and goats and are living at a poor standard of living. There are exceptional cases in which the extraction of precious metals has attracted men into the heart of deserts.

Many towns have emerged in deserts owing to the mining of gold and other precious metals. These miners and workers have to get their whole subsistence and even their drinking water from outside the region. In this way, gold mines right out in the desert have caused the growth of Kal-goorlie and Coolgardie in Australia and Cripple Creek in the United States.

In the heart of Arabia and Libya, oil has so far led to the presence of only a few groups of workers and scientists, who do not form a true self-supporting community. It is believed that at present, the desert areas are expanding and People are leaving, so the dry areas are being occupied by an increasingly smaller proportion of the world's population. Gold and petroleum have, however, made some of the regions of deserts as the areas of great attraction for labourers and technicians.

Ice Caps and Cold Regions : The second most sparsely populated are the areas of extreme cold, both in the northern and southern hemispheres. The Arctic region and the Antarctica continent are the regions of severe cold in which winters are dreaded, gloomy and long. The continent of Antarctica is almost exclusively uninhabited. In the tundra region of Arctic, the hunting and fishing peoples live, who are few in number. In fact, only a few seal-hunting Eskimos have been able to penetrate northward up to 82° north. Wherever minerals are available men have settled there to exploit them. For example, miners are extracting iron ore in Gallivare (Sweden), gold in the Yukon Valley of Canada and Fairbanks and Fort Yukon in Alaska, and gold, oil, coal and salt are precious metals in Siberia and Asiatic Tundra. For their subsistence, these miners have to depend largely on the neighbouring developed areas.

Mountainous Regions : The rugged and folded mountains are also sparsely populated. In fact, on mountains population becomes sparse with altitude. Lofty mountains are almost uninhabited. In the Himalayas, Alps, Rockies and Andese mountains, the areas above 2,500 metres are almost uninhabited. In higher latitudes, sparseness of population descends to much lower levels. In Scotland, Wales, Norway, Sweden, Russia, Canada and Greenland, sparseness begins only a few score metres above the sea. The areas of minerals are, however, exceptions. For example, in Peru and Bolivia, mining of precious metals is done above 4,000 metres. These mining places have quite substantial populations. In Kishtwar (Bhadarwah district of Jammu division), precious stones are mined at an altitude of about 5,500 metres. At the source of Indus river, the gold mines at Tok-Dschalung have helped in the development of human settlements at an altitude of 5,000 metres above the sea level.

In the tropical regions where the climate at lower altitudes is not conducive, most of the towns, cities and settlements have developed around 2000 metres above the sea level. Addis Ababa (Ethiopia), Kampala (Uganda), Quito (Ecuador), Nairobi (Kenya), Ooty (India), and Kandy (Sri Lanka) are all situated over 2000 metres above the mean sea level. Throughout the interior of Western Asia, clusters of towns are found in the heart of mountain regions. For example, Sana (2,130 m) and Taiz (2,000 m) in Yemen, Tehran (1,130 m) in Iran, Pahalgam, Gulmarg (2,600 m) and Leh (4,600 m) in India, and Lhasa (3,550 m) in Tibet are situated at a fairly high altitude.

Tropical Rain Forests : The tropical rain forests are also sparsely populated. These are the low-lying areas on both sides of equator in which rainfall and humidity remain high and the mean monthly temperatures are over 30°C throughout the year. These regions are covered by dense evergreen equatorial forests. The Amazon basin, the Congo basin, and the islands of South-east Asia, excluding

Java, are covered by such forests. The hot and humid climate, luxurious evergreen forests, infested with insects, mosquitoes, snakes, tse tse fly, etc., are least conducive for human habitation. Cultivation of crops, keeping of cattle for ranching and herding are difficult in these areas. Tropical rainforests offer some prospects for future frontier development. One advantage is the proximity to densely populated areas, it is difficult to form settlements in equatorial forested regions. At present, the Amazon and the Congo basins have a very scattered population. It is predicted that an increased population in these areas will result in a deterioration of earth's resources. In fact, often forest clearance has led to soil erosion, increased run-off, and siltation of rivers. The clearance of these forests in Thailand have adversely affected works downstreams and thus reduced the country's agricultural output. Some experts even believe that destruction of the tropical rain forests will affect the world's climatic balance by reducing the amount of moisture that is returned to the atmosphere by transpiration from luxuriant vegetation. Encroachment in tropical rain forests may also destroy the culture of hunters, gatherers and shifting cultivators. It has been evidenced in Brazil, in the destruction of these societies through violence, disease, and cultural disorientation.

All the sparsely populated areas have an irregular and sporadic type of settlement. In these regions, large areas remain uninhabited, whilst relatively small points swarm with people. The best examples of dense population are found in the larger oases of Africa and Western Asia, the islands of Java and Philippines, the isolated towns in the Congo and Amazon basins. The Himalayas and Alps have some of the large towns and resorts. It is believed that population in the sparsely populated region shall increase at a faster rate in coming decades which may create many ecological and demographic problems.

Citizens in Coming Times

The existing distribution of world population has been discussed in the preceding paras. The distribution of the world's population during the next few decades is expected to remain similar to the present pattern. The main reasons may be that the international law and politics do not permit free migration from one country to another. Secondly, the sparsely populated areas have adverse climate and generally poor in resource potential.

Prospects of large-scale expansion of population in the cold, mountainous, arid or hot and wet climates seem to be dim. The cold climates with their short growing seasons and long distances from existing commercial and industrial regions probably have limited potential for either agriculture or industry. These areas will remain inhabited by the indigenous groups that have adjusted to the harsh conditions over many centuries or by the outsiders who are busy in minerals extraction, military training and research activities.

In the arid and desert areas also, the prospects of large population agglomerations are bleak owing to the non-availability of water for drinking and irrigation purposes. Consequently, most of the deserts remain uninhabited or else thinly populated by nomads, hunters and gatherers.

The equatorial rain forests are also not very attractive regions for human habitation. Population in these areas may grow only if agriculture is developed by large-scale felling of forests. The felling of trees may be disastrous from the world's climatic point of view. The destruction of forests may destroy the climatic balance by reducing the amount of moisture that is returned to the atmosphere by transpiration from the equatorial forests.

The young folded and other rugged mountains also provide very little scope for the future expansion of human population in the hilly areas. Development of more settlements in the mountainous areas may lead to pollution, degradation of natural environment, and the excessive demands for food. Thus, it can be said safely that the future pattern of world's population distribution will remain more or less the same.

Some Popular Demographic Terms

1.	Birth Rate	Number of live births per year per 1,000 of the population.
2.	Death Rate	Number of deaths per year per 1,000 of the population.
3.	Infant Mortality Rate	Number of deaths of children below one year of age per 1,000 of live births.
4.	Life Expectancy	The average age at which people die. It is important to realize, however, that it is not the age at which most people die, i.e., in India, the figure is 69 years (2001) of age, whereas in Japan and Britain it is 80 and 77.2 years respectively; this is because more young children die in India and thus bring down the average expectancy.
5.	Natural Increase	Excess of births over deaths per 1,000 of population. This does not include increase in population due to immigration.

Formation of Age and Sex

The composition of population according to age and/or sex is known as the age and sex structure. The universal characteristics of human populations are fundamental to understanding demographic processes of fertility, mortality and migration. Age composition may be summarized in terms of age-groups (e.g., 0-15 years, 15-64 years and 65 or above). In India, the age groups are below 15, 15-59 and above 59, as the age of retirement in our country is generally 60 years. The sex ratio is most commonly expressed as the number of females per 1,000 of males.

One of the important aspects of the population study is the age composition. The age composition strongly influences the rate of growth and have profound effects on the social and economic conditions under which a population lives. Three basic determinants of age composition are: (i) natality, (ii) mortality, and (iii) mobility.

These are interdependent, and any change in one of these may influence the other two. It is through these variables that the socio-economic conditions influence the age structure. It is the fertility rate that determines the proportion of population in different age categories. That is why, most of the countries of Asia (excluding Japan, Singapore, Hong Kong and South Korea), Latin America and Africa have high fertility. Since the longevity and life expectancy in these countries is short, the proportion of population in the old age-group is also not very large. According to the estimated population of 2001, in most of the countries of Europe and North America, about 20 per cent of their total population is below the age of 15 years. In India, about 33 per cent of its population is below 15 years in age and 5.5 per cent above 60 years. On the contrary, the European countries with low fertility rate and with long expectation of life, have only 20 per cent of their populations in the young age-group. Moreover, the proportion of population in the older age-groups in developed countries is relatively large. Consequently, most of the countries of Europe, Anglo-America, Japan, Australia and New Zealand, which have almost completed their demographic transition, have an age composition in which the proportion of young population (< 15) is low and that of the old population (> 65) is high.

The age composition is also affected by the rate of mortality. In general, if the survival rate of only the childhood improves, the proportion of children will tend to rise and that of the older people will tend to fall. And, if the survival rate improvement takes place only among the older people, the proportion of older people will improve but that of children will tend to fall. Similarly, if the mortality is low both among younger and older age-groups as in the case with developed countries like Germany, Italy, Sweden, Netherlands, Belgium and Denmark, the share of the working population will be larger and the dependency ratio will be low. In contrast to this, if the decline in mortality is sharper in lower age-group than that in the older agegroup, it results in the swelling up of numbers in younger age-group as is the case with most of the populations of developing countries. According to the 2001 census, the sex ratio in India at the national level is 933 which is six points higher than that of 927 recorded at the census of 1991. The lowest sex-ratio was in Daman (591) and the highest was in Mahe (1,148) in 2001.

Migration also has a direct bearing on the age composition. Those who migrate, generally belong to the relatively younger age-group. When the youths outmigrate, the population of older adults and aged people declines after sometime. The impact of migration upon age structure is largely because of the fact that migration tends to be age-selective. Generally, people in the age-group of 15-30 years are more mobile than the people in the younger and older age-groups. Consequently, the population of the juvenile and senile age-groups increases at the place of origin of migration.

Apart from fertility, mortality and migration, the age structure of population is also influenced significantly by wars (male selective in their casuality), catastrophe, natural calamities and population policies.

Age Classes

The age structure of a given country or region may be analyzed on the basis age-groups. On the basis of physiological and economic activities, the population is generally classified into three groups: (i) the young, (ii) the adults, and (iii) the old. Although there are no standardized break points, yet the ages of 15 and 60 years are the most commonly used break Points in the developing countries and 15 and 65 years in the developed countries. The social, economic and political implications of these age-groups and the geographical variation in their distribution are worthy of serious consideration.

The Young : All over the world, the young age-group includes the population below 15 years of age. The proportion of population in this age-group in any country is determined by the stage of demographic transition through which it is passing. This proportion is large if the country is passing through the first or the second stage of demographic transition. It starts declining as the country-approaches the late stage. It is the minimum when the country is in the final stage of demographic transition. It is interesting to note that while in the world as a whole about 30 per cent population is below the age of fifteen years, the corresponding figures for the developed and the developing regions are 20 per cent and 36 per cent respectively.

In Germany, the proportion of young population (below 15 years) is only 15 per cent. In Italy, U.K. and France the adult population is only 14 per cent, 18 per cent and 17 per cent (2001) respectively. Thus, there are wide regional variations in the proportion of young population ranging from less than 18 per cent in Europe to nearly 43 per cent in Africa and about 36 per cent in Asia and Latin America. In India, about 33 per cent of the total population is below 15 years of age. These regional variations are related with the fertility patterns of different countries. Countries that are characterized with high fertility rates have large proportions of young populations and vice versa. There seems to be a direct correlation between birth rates and proportion of young population. This age-group is economically unproductive and the most expensive as it is to be provided with food, clothing, education, recreational, health and medical facilities.

The Adults : The group of adults consists of fifteen to 65 years in the developed and 15 to 60 in the developing countries. Depending upon the life expectancy, the upper limit in the developing countries is sixty and that in the developed countries is 65. The adult age-group is biologically the most reproductive, economically the most productive and demographically the most mobile (migratory). It supports the bulk of other age-groups, it carries the burden of feeding, clothing, educating the young age-group and of looking after the old-age group. The developed countries have relatively high proportion of adult population. This adult age-group is sometimes divided further into two sub-groups: (i) 16 to 34, and (ii) 35 to 59/64. The first sub-group of young adults is considered biologically most reproductive and economically more active than the second age-group of older adults.

The Old : The old age-group consists of the people who have achieved the age of 60 years in the developing countries and 65 or above in the developed countries. Such people are called as the senior citizens. The proportion of people in this age-group, by and large, increases as the population of a country completes its demographic evolution. In the developed countries, the number of females in this age-group is more than that of the males because the proportion of population in this age-group is governed largely by mortality rates, and since the male mortality rate in the developed countries is higher than that of the females at all ages, the females tend to outnumber the males in this age-group. The males belonging to this age-group usually remain productive and biologically less reproductive. This age-group is largely an economic burden upon the adult age-group as it is to be provided with food, clothing and adequate health care under the social security system.

Literacy : Literacy has been defined as the ability to read and write with under-standing. Literacy reflects the socio-economic and cultural set-up of a nation, ethnic group or community. Literacy is essential not only for the eradication of poverty, but also for mental isolation for cultivating peaceful and friendly international relations and for permitting the free play of demographic processes as well.

The concept of literacy, which varies from country to country, generally refers to the minimum level of literacy skills. This minimum level of skills varies from ability to communicate orally, to make a check of a variety of difficult arithmetical computations. The length of schooling has, however, often been considered as a basis of distinguishing between a literate and illiterate. Trewartha (1969) considers the basis of length of schooling not a valid measurement of educational accomplishments. He also disapproves of the ability to read and write one's name in the language of his country as the criterion for defining a literate. In 1930, Finland applied perhaps the most strict definition where only those persons were classified as literate who passed a rather difficult test.

Those who failed were divided into two categories: (i) the semi-literate persons, who could read and write but made orthographic errors; and (ii) the illiterates, who could neither read nor write (UNESCO, 1957). The Population Commission of the United Nations considers the ability to both read and write a simple message with understanding in any language as a sufficient basis for classifying a persons as literate. The Indian census has adopted this definition and now many of the developing countries are shifting to this definition too. However, a distinction can be made between the literates and the educated as has been done in the case of India. All those persons who are classified as literates on the basis of their ability to both read and write are further sub-divided into a number of categories on the basis of their length of schooling.

There are inter-regional and intra-regional variations in the literacy rate. The literacy variations are quite significant between the developed and the developing countries. In general, low literacy rates are associated with high birth rates. If a population is illiterate, it will tend to change slowly and to resist new ideas and innovations. Illiteracy is a form of isolation. Illiteracy takes away from man his dignity, perpetuates ignorance, poverty, mental isolation, and friendly international relations and thus hampers socio-economic and cultural advancement. It is not surprising, therefore, that the countries with the high birth rates have high proportion of illiterate population.

Scene of Literacy

The literacy rate in a country or region is determined largely by the historical, social, cultural, political and economic factors. The main factors that determine the literacy are: (i) cost of education, (ii) political/ideological background, (iii) type of economy, (iv) standard of living, (v) degree of ur-banization, (vi) stage of technological advancement, (vii) degree of development of means of transportation and communication, (viii) religious background, (ix) medium of instruction, (x) status of women in the society, (xi) prejudices against the females' mobility and education, (xii) availability of educational institutions, (xiii) general value system, and (xiv) public policies.

The political situation and administrative governance have affected the rate of literacy in the ex-colonial countries. In general, the literacy rate in the ex-colonies has been generally low. The present governments in these countries have taken the literacy drive, but the scarcity of resources is coming in the way of eradication of illiteracy.

In the second part of the 20th century, the areas of isolation and relative isolation have been connected by metalled roads. The development of transport network and improved accessibility have increased the spatial interaction. The so-called rural isolation has been broken. The educational institutions in the urban areas now have become accessible to the country-side. The rural masses of the developing countries are now increasingly appreciating the importance of education and literacy.

In the propagation of education, the role of religion is also quite vital. Similarly, language in which the education is usually imparted is another important determinant of literacy rates of an area. The mother tongue as a Medium of instruction can help to accentuate the literacy rate.

Discrimination and prejudice against females also affect adversely the literacy rate in a region. Those societies which do not give equal status to males and females suffer from low female literacy rates. Similarly, the societies where the females are not allowed mobility or which suffer from the prejudice against females' education also display low female literacy and hence low general literacy. It is perhaps one of the reasons why the female literacy rate among the Muslim women is one of the lowest. The availability of educational institutions is also largely responsible for the enhancement of literacy rate.

The literacy rate is also closely influenced by the type of economy. The differences in the literacy levels of the industrial and agricultural nations contrast so much that one cannot help inferring a correlation between the type of economy and the literacy rates. The requirements of the non-agricultural economy are such that acquisition of literacy skills becomes a functional prerequisite. On the other hand, the agricultural operations, especially in the developing countries, are such that these do not have any demand on education.

The standard of living also affects the literacy rate. In general, there is a positive correlation between the literacy rates and the standard of living. Such a correlation has greater significance for poor countries where the appalling poverty of masses demands different priorities. The factor of cost of education is also important as the poor people cannot afford the high cost of education. It is difficult to expect the children belonging to the families that lie below the poverty line to go to schools. The experience of India in this regard reveals that even if the education is made compulsory and free, the extremely poor families prefer their children to help them in making them an earning, howsoever, meagre it may, be, rather than sparing them for school.

The level of urbanization is also an important determinant of literacy. There is a positive correlation between literacy rate and the degree of urban development. Thus, the most urbanized societies have a very high or even universal literacy rates, while the rural and agrarian societies have low level of literacy rates.

Spatial Pattern of Literacy : Adequate and reliable data on spatial pattern of literacy are not available. Even the United Nations has not prepared a map showing the spatial distribution of literacy in the world. However, on the basis of the available data, it may be said that the continent of Africa has the lowest literacy rate. The females are particularly more illiterate in the developing countries. Next to Africa comes the continent of Asia where also in most of the countries the literacy rate is below 30 per cent. The female literacy in Asia is also relatively low. There is a marked variation in the rural and urban literacy also. The literacy rate in the Latin American countries is relatively higher. The continents of Europe, North America and Oceania have almost universal literacy rates.

In India, according to the census of 2001, the literacy rate was 65.49 per cent. This, however, does not include the children who are below six years of age. The highest literacy rate being 96.64 per cent was in Aizawl (Mizoram), while the lowest literacy rate was in Dantewada (Chhattisgarh) being 30.01 per cent. The male and female literacy rates were 75.85 and 54.16 per cent respectively.

India is one of the least literate societies in the world. India's seemingly large pool of technically educated and English speaking manpower is deceptive and presents a misleading impression of the levels of education and literacy. The decadal difference in literacy rates between 1991 and 2001 was 13.29 per cent. High literacy rates prevail in Kerala, Mizoram, Goa, Delhi and a few union territories. However, even in 2001, literacy rates in the states of Bihar and Uttar Pradesh were below 55 per cent. Moreover, there are also gender disparities in literacy rate. In 2001, the lowest literacy rate was recorded in the state of Bihar where it was only 47.53 per cent.

The quality of education provided in India so far leaves much to be desired. For example, most people educated upto primary level revert to illiteracy soon after they drop out of school. Sustained efforts thus need to be made to improve the quality of education, besides increasing overall levels of literacy.

The relatively low literacy rate in India may be attributed to the rigid caste system of Hindus, and the religious orthodoxy of the Muslims. Among the Hindus, the Sudras (lower castes) were denied education as they were considered to be untouchables. Moreover, the factors of poverty, unemployment, early marriage, and prejudice against their mobility have kept the females of India far behind their male counterparts in the matter of literacy and education. The rural and urban differentials in the literacy rates of India emanate from the differences in the type of economy, social life and migration pattern of the two areas.

Concepts of Citizenship

Right from the rise of human civilization, the scholars did not agree on the population growth. While some of them argued a continuous increase in population for the survival of human species, races and nations, the others believed that the fast growth of population had been the cause of disappearance of many civilizations from the earth surface. It is, therefore, quite appropriate to give a historical overview of population growth.

The Chinese philosopher Confucious argued that a numerical balance be maintained between population and environment. Thus, he was not in favour of unchecked growth of population. In fact, he was the first who gave the concept of optimum population level. In ancient Greece, the earliest thinkers favoured the expansion of population, but Plato was a restrictionist who advocated as absolute limit of population. Aristotle and Polybius attributed the fall of Hellenic civilization to the decline in population growth.

Early Rome was characterized by a fertility cult. Growth of population they believed was necessary for military and political expansion. They were of the opinion that the primary function of marriage was to provide citizens and soldiers for the state. It was in this background that Augustus introduced legislation to encourage population growth. In spite of these efforts, Rome's lack of men finally contributed to its downfall, since it was unable to hold back the barbarian invasions.

Medieval period was the period of invasions, battles and wars, which gave rise to the feeling that population growth is necessary to survive and to defend. It was a common belief that who will carry the weapons, if men are lacking? Man was considered as the best wealth. Most orthodox thinkers in the 18th century were of the opinion that people and beasts must be multiplied. The German scholars also attributed the power of states mainly to population. The first thinker who took the question of population in a formal way was Malthus Thomas.

Malthusian Theory of Population Growth : Malthus Thomas Robert, the English economist and demographer, is well-known for his theory of population growth. He published An Essay on Principle of Population in 1798 and it is to his ideas that the subsequent thinking on the economic approach to demography may be traced. His general view was that population tends to increase faster than the means of subsistence. The fast increase in population absorbs all economic gains, unless controlled by what he termed 'preventive' and 'positive' checks. He maintained that population if unchecked tended to increase at a geometric rate (i.e., 1, 2, 4, 8, 16...) while subsistence increased at an arithmetic rate (1, 2, 3, 4, 5...). The population doubles after every twenty-five years, and in two centuries the population would be to the means of subsistence as 256 to 9, in the three centuries as 4096 to 13 and in two thousand years the difference would be almost incalculable. Malthus principle states that:

1. Population is necessarily limited by the means of subsistence.
2. Population invariably increases where the means of subsistence increased, unless prevented by some very powerful and obvious checks.
3. These checks, and the checks which repress the superior power of population and keep its effects on a level with the means of subsistence, all are resolvable into moral restraint, vice and misery.

Malthus believed that two of man's characteristics essential to the maintenance of life were immutable, were laws, and were antagonistic: (i) the need for food, and (ii) the passion between the sexes. It was the second which led people to marry at a relatively early age and would result in such a large number of births that the population would double itself in few years if unchecked by misery and vice.

The widening gap between population and subsistence will increase a man's tendency to press upon the mean's of subsistence. With the result, the society gets divided into two sets of people—the rich (haves) and the poor (have nots)—giving rise to capitalistic set-up. The rich who are the owners of the means of production earn profit and accumulate capital. With the increased capital, they enhance their consumption but do not increase their population for fear of decline in their standard of living. Malthus defends the capitalistic set-up of society on the ground that if the capital was to be distributed among poor, it will not be available for investment on the mode of production. Thus, the rich will continue to grow richer and the poor constituting the labour class poorer. In his opinion, the increasing gap between the population and resources shall ultimately lead to the point where misery and poverty shall become inevitable. According to him, postponement of marriage was and would continue to be the chief preventive. He did believe that the preventive checks, such as delayed marriage, moral restraint, etc., would so reduce man's rate of production that the positive checks would not operate continuously. He believed that change in society and social institutions would mitigate the positive checks for a shortwhile, until man's number would again grow more under his new institutions, and that man would then suffer from such positive checks like vicious customs about women, pestilence, war, hunger, disease, etc.

The 'positive' and 'preventive' checks which occur in human populations to prevent excessive growth relate to practices affecting mortality and fertility respectively. So his 'positive' checks included wars, disease, poverty, and especially lack of food. His 'preventive' checks included principally 'moral restraint', or the postponement of marriage, and 'vice' in which he included adultery, birth control and abortion. He saw the tension between population and resources as a major cause of the misery of much of the humanity. Malthus was not, however, in favour of contraceptive methods, since their use did not generate the same drive to work hard as would a postponement of marriage. He stressed the negative correlation between station in life and number of children and, in order to induce in the lower classes the self-control and social responsibility which he saw in the middle classes, Malthus asserted that the poor should be better paid and educated.

The Malthusian theory has been criticized on several counts. His thesis that population was growing quickly and that man was a biological as well as social being, depending on sexual drive and food. Yet, he confused moralist and scientific approaches. Marx was one of the most powerful critics of Malthus asserting that poverty is the result of unjust social institution of capitalism rather than of population growth. No country of Europe or North America except possibly Ireland conformed to the Malthusian prediction. It has also been argued that Malthus' reactionary views impeded the development of demography as a science.

The main points of criticism of Malthusian theory have been given as under:

1. The basic assumption of Malthus on passion between the sexes has been questioned on the ground that the desire to have children cannot be mixed up with passion and desire for sex. The desire for sex is biological instinct, whereas the desire to have children is a social instinct.
2. The validity of his two sets of ratios has also been questioned by hiscritics. Population has rarely grown in geometrical proportion and means of production have rarely multiplied in arithmetic progression.

3. The span of twenty-five years assumed by Malthus to allow population to double itself also does not seem to be anywhere near reality. The doubling period for a population varies from country to country and from region to region, depending upon the stage of its economy, and scientific and technological advancement. For example, it took only about twenty nine years for the population of Bangladesh, Iran and Mexico to double itself and over 280 years in United Kingdom, 270 years in Japan, 100 years in United States while the doubling period of population in Sweden, Denmark, Finland and Norway may be more than 300 years and it is projected that the population of Russia and Germany at the present rate will not double even during the infinite period of time.
4. Malthus overemphasized the 'positive' checks and did not visualize the role of 'preventive' checks like contraceptives and family planning.
5. Moreover, natural calamities have occurred in underpopulated areas also and thus there was no causal relationship between positive checks and over-population.
6. Malthus has been severely criticized for ignoring the role of changing technology and the consequent transformation in socio-economic set-up of a society.
7. Malthus also failed to realize even the biological limitations that a population cannot grow beyond a certain limit.

Malthus was not so much wrong as he was premature. He recognized that migration and improved techniques of production would temporarily postpone the difficulties engendered by population increase. But he could not have been expected to foresee the tremendous burst of productivity in the modern world (especially in Europe and America) during the 19th and 20th centuries which brought progressive release from the positive checks.

In spite of all these criticisms, Malthusian principle of population has been successful in highlighting the urgency to maintaining a balanced relationship between population and means of subsistence. The critics of Malthus failed to realize that it was because of a large measure of truth in Malthusian principle of population that men of today feel the need of resorting to contraception to keep their families within reasonable limits. Another main contribution of Malthus was to bring the study of population into the fold of social sciences. It gave a new line of thinking whereby the dynamics of population growth were viewed in the context of man's welfare. Above all, the Malthusian principle of population initiated theory building and for this reason, his work is of great value.

Marxian Theory of Population Growth : Marx was a creative thinker who made a scientific interpretation of human history. He argued that like scientific and universal laws in physical sciences, there are universal laws of social processes. According to him, the essence of history is change in modes of production in any society; and this change is always progressive. For example, from the stage of nomads, mankind turned to the stage of settled living. Hunting, fruit gathering and sheep rearing gave way to domesticating animals and private agriculture. Then came urban culture and rich diversity of vocations. Population increased, inhibiting factors weakened, private property emerged, and a stage was reached when private means of production could also be refined in a hundred ways. This led to the formation of economic classes, and there occurred a social hierarchy, codes of behaviour, rules of punishment for crime and misconduct. Subsequently, the Industrial Revolution (1779) gave birth to the capitalism and thereby emerged the classes of haves and have nots. In Marxian thought, the division of people into classes had the effect of initiation of class struggle as the class interests were mutually hostile and irreconcilable. In brief, there was a class of employees and another of the employed. In familiar Marx terminology, this would be called as exploiters versus exploited.

It was in this backdrop that Karl Heinrich Marx (1818-1883), while postulating his general theory of communism and scientific interpretation of history, gave some ideas about population growth. As a strong believer in dialectical materialism, he considers that society, especially the feudal and capitalist society, has two major economic classes, viz., the rich and the poor. The rich were those who possess the modes of production (land and factories etc.) and they earned profit by exploiting the labourers and the environment (resources). The owner of modes of production has the freedom to offer employment and the workers, being too poor, cannot resist the terms the employer offers. Thus, the workers by contract sell their energy, skill and will to work in return for agreed wages. Under this system, the owner of means of production retains the right to terminate the contract of employment at his sole discretion. He hires and fires at will. In most cases, the law is on his side. Where this is not so, law courts can be induced to prove helpful to this cause. Moreover, the owner is the sole judge of the utility and performance of his workers.

This has been the situation ever since man started employing other men for the production of goods in the field, the shop, the household and the factory. The employer earns profit by exploiting the workers as they are underpaid in the form of wages. This profit was termed as 'surplus value' by Marx. Surplus value was the positive difference between the exchange value and subsistence value (wages). The profit earned by the rich led to the process of capital accumulation. In consequence to this, the poor workers (labour class) try to accumulate labour, the only commodity they possess, through rapid population growth (Marx, 1967: 218). Improvements in the technology of production and rapid growth of population among the poor workers lead to surplus labour and unemployment. This is the main cause of misery, unemployment, undernourishment and poverty in the society. The population of the poor grows at a faster pace as compared to that of the small proportion of population who possess the modes of production (the rich and ultra-rich).

Fundamental change in history, according to Marx, is on account of class struggle in human societies. This is also the crucial key to basic understanding of population growth. Judged in this light, not only can the past pattern of growth of population can be interpreted, but even the future trends of population growth can also be reasonably projected.

Marx further advocated that it is the working population which, while affecting the accumulation of capital, also produced the means whereby itself is rendered relatively superfluous, which is turned into a relatively surplus population. This process leads to unemployment and poverty in the society. With the passage of time there occur new innovations in the methods of production, the improved technology render the growing population of poor people unemployed, which sharpen and intensify class struggle. For this reason, class struggle and poverty are seen as the main driving forces of rapid population growth among poor workers.

Marxian theory of population is based on as a reaction to the capitalistic mode of production and governance. According to Marx, poverty and misery were not natural inevitabilities but unpleasant gifts of capitalism. Misery, poverty, unemployment and fast growth of population can disappear if capitalistic form of social order is replaced by communism. He had a convinced opinion that poverty and misery are the outcomes of unemployment and underemployment arising out of inability of the capitalistic social order to provide jobs to all, regardless of the speed with which the population increased.

Marx was not in agreement with Malthus who postulated that population tends to increase faster than the means of subsistence, thus absorbing all economic gains, unless controlled by what he termed 'preventive' and 'positive' checks.

Marx places the workers (masses) at the centre of main historic process on which depend the growth of population and the course of history. He also asserted that on the top is, and has been, the ruling class, small in numbers but unchallenged in authority and power; and down below is that large body of the masses which lives in utter misery and is content to be exploit and driven like dumb creatures.

In the opinion of Marx, there could be no one universal law (natural law) of population growth. The growth pattern and dynamics of population change with the change in mode of production. Mode of production is a specific set of forces of production (labour, capital, material, machinery, etc.) patterned into a specific set of relations of production developed as a result of agreement between the employer and the workers.

According to Marx, each mode of production had its own economic and demographic laws. As stated above, the demographic characteristics and growth patterns of population change with the change in mode of production. In other words, the growth pattern of population of a nomadic society will be different from that of settled cultivators and the demographic attributes of an industrial urban society will be different from that of an agrarian rural society.

Thus, in the opinion of Marx, economic classes and private property are the principal twin evils which lead to poverty, unemployment and a growth of population. The private property hurts the general good of the community at large. In particular, property that qualifies as primary means of production such as mines, land, agricultural fields, orchards, factories mills etc., cannot be permitted to be owned privately. These resources should be owned by the community at large, if the rapid growth of population is to be checked and poverty and miseries are to be eliminated from the society. Marx had full faith in the ability of communist method of production to give full employment and a good living to all able-bodied worker regardless of the rate of increase in their number.

One basic postulate of Marx about the growth of population was that in the capitalistic form of society the supply of labourer increases much faster than the opportunities of employment. This surplus population becomes an industrial reserve army of unemployed and underemployed hands. The movement of wage levels is determined by the magnitude of working population among this industrial reserve army. This critical proportion workers among reserve army is controlled by expansion or contraction of capital. The birth and death rates as well as size of family in turn have inverse correlation with the level of wages, i.e., means of subsistence at the disposal of different categories of workers. Such a class of workers which is more prone to become a part of reserve army or surplus population shall have lower wage level and hence high birth and death rates, a situation found in most of the underdeveloped and developing countries of Africa and Asia.

Marx, in his theory of population, tried to establish a relationship between capital accumulation, labour demand, surplus population (unemployed and underemployed workers), wage levels, standard of living, poverty and rates of fertility and mortality and growth of population. In his opinion all these are closely related in the capitalistic form of society in which means of production are owned by a small proportion of population and the rest are the workers who are being exploited by the employers.

Marx postulation about population exposes the weakness of the capitalistic form of social order in which individuals have the right to accumulate huge amount of wealth and in which labourers are exploited by giving them low wages, making them poor and helping them increasing population at a faster pace. This theory has, however, been criticized at several counts, as under:

1. The unprecedented growth of population in the world is not due to the lower wages, unemployment and underemployment but mainly owing to the extension in the medical facilities and health care services which substantially reduced the death rates without arresting the birth rates.
2. The increase in the population does not automatically lead to the decline in real wages as has been argued by Marx. There are numerous socio-political and economic factors which determine the wage levels and employment opportunities in a society.
3. Marx attempted to establish a positive correlation between the levels of wages and the birth rate of population growth, i.e., higher the wages, lower the birth rate. The faith and religion of the population has, however, not been taken into consideration by Marx in his postulate. There are numerous affluent ethnic and religious groups in the developed and developing countries in which the rate of birth is significantly high.
4. Marx overemphasized and considered private property as the main cause of all evils including poverty, misery, unemployment and fast growth of population. The social norms, education level, technological advancements and attitude towards family are all vital determinants of population growth. Thus private property may not be blamed as the sole factor for population increase.
5. Marx theory of population growth may be relevant to the capitalistic societies and in all probability would not operate in a feudalistic, socialistic and primitive hunting and food gathering societies.

6. The physical environment (terrain, climate, soil fertility, etc.) as the determinant of birth and death rates has been given adequate weight by Marx in the growth of population. There are scholars like Spencer who believed that man had no control on his reproduction capacity as the forces of evolution are quite beyond his control.
7. There are demographers like Dumont who believed in the prevalence of social capillarity where the need for smaller families would be generated by the desire for better economic status.

Despite all these criticisms, it can be said that after Malthus, Karl Marx attempted to give a scientific explanation of the growth of population which was based largely on the information available from the capitalist countries. Growth of population in a region is, however, controlled by the physical, socio-cultural and economic conditions. It is because of these factors that a universal model of population growth cannot be postulated. The Demographic Transition Theory, developed in the 20th century, gives a more cogent, logical and systematic explanation of population growth.

Demographic Transition Theory : The demographic transition theory is one of the most important population theories which is the best documented by the data and statistics of recent demographic history. In its original form, the demographic transition theory was put forward by W.S. Thompson (1929) and Frank W. Notestein (1945). These scholars based their statements and arguments on the trends in fertility and mortality, being experienced in Europe, North America and Australia.

The theory postulates a particular pattern of demographic change from a high fertility and high mortality to a low fertility and low mortality when a society progresses from a largely rural agrarian and illiterate society to a dominantly urban, industrial, literate and modern society. The three very clearly stated hypotheses involved in the process are:

(i) that the decline in mortality comes before the decline in fertility;

(ii) that the fertility eventually declines to match mortality; and

(iii) that socio-economic transformation of a society takes place simultaneously with its demographic transformation.

In the present-day world, as would be true of any point in time, different countries of the world are at different stages of the demographic transition. In the opinion of Trewartha, this is largely due to the dual nature of man. According to him, biologically, man is same everywhere and is engaged in the process of reproduction but culturally man differs from one part of the world to another. It is the cultural diversity of man that gives rise to varying fertility patterns in different areas resulting in different stages of demographic transition. The demographic transition theory is characterized by conspicuous transition stages. The transition from high birth and death rates to low rates can be divided into following five stages.

Stage I	High and fluctuating birth and death rates, and slow population growth.
Stage II	High birth rates and declining death rates and rapid population growth.
Stage III	Declining birth rates and low death rates, and declining rate of population growth.
Stage IV	Low birth and death rates, and slow population growth.
Stage V	Birth and death rates approximately equal, which in time will result in zero population growth.

In the first stage, the fertility is over 35 per thousand and the mortality is also high being more than 35 per thousand. The behaviour of mortality is, however, erratic due to epidemics and variable food supply. This stage, thus, postulates a stable and slowly growing population where the people are engaged in wasteful process of production. This stage mainly occurs in agrarian societies where the population densities are low or moderate, generally productivity level is low, large-sized families are an asset, life expectancy in low, the development of agricultural sector is at its infancy stage, masses are illiterate, technological know-how is lacking and urban development is limited. About two hundred years ago, all the countries of the world were at this stage of demographic transition. At present, it may be difficult to ascertain whether any country in the world would still be at this initial stage of demographic transition because the data pertaining to fertility and mortality for such areas would either be lacking or would not be reliable.

Moreover, the diffusion of modern technology has also been so fast particularly in the field of medicine, that it is very difficult to find a solitary example of a country which may still be unaffected by the mortality declines taking place all over the world. It is in this context that the first stage has been called as the pre-industrial and the pre-modern stage.

The second stage of demographic transition is characterized by a high and gradual declining fertility of over thirty per thousand and a sharply reduced mortality rate of over fifteen per thousand. In this expanding stage of demographic transition, while the improvements in health and sanitation conditions result in sharp declines in the mortality rates, the fertility maintains a high level, at least in the early second stage. As the second stage prolongs, the fertility also shows signs of gradual decline.

A distinction has often been made between the early second stage with high fertility and declining mortality and the late second stage with slowly declining fertility and sharply declining mortality.

In the second stage, as a whole, the population expands, firstly, at a gradual increasing rate and afterwards at a gradual subsiding rate. In the wake of population explosion associated with the widening gap between the two vital rates, the population of resource mobilization becomes significant. The life expectancy starts improving. The processes of industrialization, urbanization, and modernization become prominent.

The large families are no longer an asset. Consequently, the fertility undergoes a gradual decline leading to a gradual squeeze of rate of natural increase at the tail end of the second stage. Most of the less developed countries of the world are passing through this explosive stage of demographic transition because of wide-spread penetration of modern medicines and sanitation measures have drastically reduced their mortality rates whereas their fertility rates are high. The countries like India, Pakistan, Bangladesh, Nepal, Iran, Yemen, Kenya and Indonesia belong to the late second stage where the fertility rates have started declining gradually but since the decline in mortality rates has been sharper there is tremendous growth of population.

The third, or the late expanding, phase is characterized by a slowing in the growth rate as the death rate stabilized at a low level and the birth rate declines. This decline is associated with the growth of an urban/industrial society.

In the last stage of demographic transition, both birth and death rates decline appreciably. The population is either stable or grows only slowly. In this stage, the population is highly industrialized and urbanized. The technical know-how is abundant, the deliberate controls on family size are common, the literacy and education levels are high, and the degree of labour specialization is also very high. Anglo-America, West

European countries, Australia, New Zealand, Singapore, Hong Kong and Japan are supposed to have reached this stage of demographic transition.

Although the theory of demographic transition has been appreciated widely by the demographers, it has been criticized on many counts. This theory is based upon empirical observations in Europe, America and Australia. Loschky and Wildcose asserted that the theory is neither predictive nor its stages are sequential and inevitable. For example, China has entered the third stage of demographic transition owing to the one child policy adopted by the government in the eighties. The role of man's technical innovations cannot be underrated, particularly in the field of medicine which can arrest the rate of mortality.

In spite of these shortcomings, the theory does provide an effective portrayal of the world's demographic history at macro level of generalization. From such a model one should not expect that there would be a typical period of time for transition, a typical sequential of fertility and mortality patterns and a typical socio-economic fabric of each stage. As an empirical generalization developed on the basis of observing the demographic trend in the West, the transition model does help in understanding the transition process for any country provided the situational context is properly understood. It would be academically unsound to expect all the countries of the world to follow the same sequential pattern as experienced by the European countries, because the present context is a significant deviation from the 19th century European context.

Best Denizens

The distributional pattern and density of population are highly unevenly distributed over the earth surface and same is the case with natural resources. The pressure of population on the resource base and the available technology determine whether the region is overpopulated or underpopulated. In other words, the extent to which resources are utilized and the way in which they are used determine whether an area is underpopulated or over-populated. A country is said to have an optimum population when the number of people is in balance with the available resources. Optimum conditions can only be maintained if the exploitation of new resources or the development of other forms of employment keeps pace with increases in population. If the population becomes too large, the law of diminishing returns begins to operate. Up to a certain point an increase in the number of people working on the land leads to a marked increase in production. Once the optimum population has been reached, however, a further increase may increase production but at a decreasing rate, so that output per capita declines. As more people become dependent on the same resource base, each individual will became poorer.

Contrary to this, if there are not enough people to develop all the resources of an area, its standard of living may remain lower than it could be, were its full potential realized. Thus, Brazil, with about 14 people per sq km may be considered underdeveloped today. Yet, before it was colonized by the Europeans, it may not have been underpopulated even though there were fewer people, because the range of resources utilized by the indigenous population was also far smaller than the resource base exploited today. Similarly, in terms of present-day technology, Central Asia is considered underpopulated. There are vast mineral resources which could support industrial development in Kazakhistan, Uzbekistan, Turkmenistan and Russian Siberia. But, in the past, Central Asia was inhabited mainly by pastoralists who knew nothing of modern technology. The resources which they were capable of exploiting were often overstrained. Consequently, during the medieval period, waves of Central Asian people outmigrated and invaded the neighbouring and distant areas in search of food and land,

and spread as far afield as Eastern Europe, South-west Asia, South Asia and China.

Under population or overpopulation, therefore, must be considered mainly in terms of technology and economic stage of development of the country/region concerned, and the standard by which this is measured in that of the industrial countries. An advanced country can be considered as one where agriculture, industry, communications, trade and commerce, and social services are all well-developed and the resources of the country are fully and judiciously utilized.

Issues of Denizens

The growing pressure of population on resource base, especially on arable land, has created many socio-economic, cultural, political ecological and environmental problems. The population problems vary in space and time and differ from region to region. These problems may be more systematically examined if we take the problems of the developed and developing countries separately.

Problems of Developing Countries : Most of the world population lives in the developing countries. China and India support over 21 per cent and about 17 per cent of the total world population respectively. Taking together the developing countries have over three-fourth of the total world population. The level of technological development in these countries is, however, low which is affecting the agricultural efficiency and coming in the way of industrial development, despite the availability of local resources. India, Pakistan, China, Brazil, Bangladesh, Myanmar (Burma), Nepal, Indonesia, Malaysia, Vietnam, Thailand, Maldive, Philippines, and most of the African countries are such of the developing countries. There are many countries which are underdeveloped because they have small and inadequate population (workforce) to utilize their abundant resources. Such countries include Brazil,

Columbia, Peru, Zaire, Russian Siberia, Saudi Arabia, Kazakhstan, Uzbekistan, Turkmenis-tan, Kirgyztan and Tajakistan. These countries have tremendous resources which cannot be developed because of lack of population. Their problems are often accentuated by adverse climatic conditions. Rapid growth of population, unemployment, inadequacy of housing and health, underutilization of resources and slow growth of industries are their main problems. Some of the important population problems of the developing countries have been briefly examined in the following paras:

Rapid Growth of Population: In most of the developing countries, the birth rate is high as the death rate has been checked because of the development and extension of medical facilities. Moreover, family planning in most of these countries is not practised sincerely on a large scale. This situation has resulted into large proportion of young people who are depêndent on relatively small workforce of the population. For example, in India and Pakistan, about 33 and 42 per cent of their population respectively is below 15 years of age. This large proportion of young population puts great pressure on the available medical, educational and other social amenities.

Unemployment: In most of the developing countries, the population is largely dependent on agriculture. The secondary and tertiary sectors (industries and services) are relatively less developed. There are very limited opportunities for the semi-skilled, unskilled and highly educated people. The rural areas are the places wherefrom large number of unskilled workers face the problem of unemployment. The educated and skilled technocrats also have very limited opportunities of employment. Consequently, both the educated and uneducated, skilled and unskilled workers try to emigrate to other countries in search of employment. Those who find it difficult, migrate to big towns where it is often even more difficult to find employment. Moreover, the towns become overcrowded, making living conditions poorer, and resulting into socio-economic and environmental problems.

Poor Standard of Living and Malnutrition: There is shortage of nourishment, especially that of balanced diet in the developing countries. The standard of living is low and housing conditions are often poor. The standard of hygiene and quality of nutrition are also low, which lead to health Problems such as deficiency diseases. The ignorance of people, the inadequacies of medical facilities, and lack of financial resources come in the way of improving the housing and health conditions.

Mismanagement of Agricultural Resources: By and large, most of the developing and underdeveloped countries have agrarian economy. The agriculture is mostly done by traditional methods, obsolete equipments and inadequate financial resources. Owing to the lack of funds and finances, the farmers are unable to apply chemical fertilizers and other inputs in required quantities. Consequently, the production per unit area is low. The fragmentation and small size of holdings and land tenancy systems are also some of the serious barriers in the modernization of agriculture. In such countries, land, their ultimate asset, is thus either under utilized or mis-utilized. Many of the farmers, being tradition bound, do not accept the innovations and new ideas. Consequently, their agricultural techniques remain traditional and production much below than their potential.

Slow Growth of Industrial Sector : In most of the underdeveloped and developing countries, the industrial sector is generally not very strong. There is lack of local capital which makes the actual exploitation of resources or setting of factories difficult. The workforce, though large in number, is generally unskilled and has no background of industrial development. Similarly, although the large population should provide a good market for the finished goods, the majority of the people are poor and can not afford to buy the products. The poor industrial base, lack of capital and poverty of people create a vicious circle and the growth of industries is hampered. The pressure on agricultural resources continuously increases.

Orthodoxy: As said earlier, the people in the underdeveloped and developing countries are tradition bound, and less exposed to the outside world. Moreover, they are religious in their attitudes who do not accept easily new ideas and modern style of life. Being religious, they generally do not observe family planning. Birth control is forbidden by the Catholic Church. In India, caste restrictions on occupation also help to slow down transformation of society and process of development. For the removal of such attitudes and for the eradication of blind faith and orthodoxy, large-scale literacy and mass education are necessary.

Problems of Under Population: Some of the underdeveloped countries/regions are underpopulated. The population of such countries are different from those of the densely populated countries. In the sparsely populated countries, the growth of population, despite high birth rates, is slow. Immigration is an important source of population but it is usually to the towns. At the same time, the towns with their better social amenities attract people from the already sparsely settled countryside. Imbalance between town and country is a major problem of underpopulated countries.

(a) Being the areas of isolation and relative isolation, it is difficult to increase settlement in sparsely populated areas because people are generally unwilling to forego the amenities of the town. The government of the erstwhile Soviet Union provided lucrative incentives in the form of heavy amounts for the construction of houses in Siberia but people remained reluctant. In areas of sparse population, it is expensive and uneconomic to provide elaborate communications, health, education and other social amenities.

(b) In the sparsely populated areas, a judicious utilization of resources is not possible. Agricultural resources are more difficult to develop because they require more and harder work over a long period of years before they show a good return. The sparsely populated countries also have slow growth of industries. There is generally a shortage of skilled labour as in the case of African, Latin American and many of the South-west, Central and South-east Asian countries. Where skilled labour is to be brought from other countries, the cost of the production of industrial goods goes up. Moreover, the small population does not provide a good market even where the standard of living is high.

(c) Many underpopulated countries have hostile climatic or terrain and topographical conditions. These hostile conditions make settlement difficult or dangerous for immigrants. Such conditions obstruct development. To open up underpopulated areas is both difficult and very expensive. A sound immigration policy, however, can help in the rational utilization of resources of underpopulated regions.

Problems of Developed Countries : The developed countries are highly industrialized and urbanized. In these countries, the per capita income is not only high, most of their population is dependent either on secondary or tertiary sector. These countries, despite high degree of development, efficient agriculture and large-scale industrial production, are also confronted with many of the population problems.

Long Span of Life: In the developed countries, the birth and death rates are low. Moreover, the proportion of younger people in the population is relatively small and the low death rate and high life expectancy mean that there is an ever-increasing proportion of older people in the population. For example in Germany, Italy, U.K. and Belgium about 18 per cent of their population is over 65 years of age. These retired people largely remain dependent on the working population.

Small Work Force: As standard of education improves, children remain longer at school and join the work force late. This, combined with low birth rate, means that the labour force expands only slowly while industrial and other employment opportunities continue to multiply. Despite a high degree of mechanization in most industries many countries are short of workers. In Europe for, example, Germany, France, Netherlands, Switzerland and Russia are short of workers. Another problem is that work force is generally well-educated and skilled and there is a serious shortage of unskilled workers. Because the majority of workers are skilled and the work force is relatively small, wages of unskilled workers are very high. For example, a coal miner in U.K. and Germany gets as high a salary as that of a university professor. Moreover, the coal miners have to work only three days or 24 hours per week.

Rural Population: There are more social amenities like universities, colleges, hospitals, banks and places of recreation in the urban areas. For this reason, the rural youth outmigrate from their villages and start their career in towns and cities. The fewer, especially the old people, live in villages. The agricultural sector suffers adversely because of the non-availability of work-force. The rural areas get depopulated and consequently, less social amenities are provided in the countryside. The standard of living of the villages suffers a decline.

Urbanization: As town expands, the pressure on transport, water supplies, sewage and refuse disposal grows and creates problems. Smoke and chemical effluents from factories produce air and water pollution. Traffic congestion and noise are other problems. Tension created by urban life lead to a far higher incidence of mental illness, heart troubles, breathing problems and madness in developed countries than that in underdeveloped ones.

Urban sprawl and slums expansion in some of the developed countries e.g., U.S.A., are the other major issues which, create many social and environmental problems. The highly productive agricultural land is encroached upon by urban houses, roads and industries, resulting into a decline in arable land.

Thus, the underdeveloped, developing and developed countries have some problems in common. Most of the developed countries have areas where agriculture or industry could be improved or where the population is too large and dense. Similarly, the developing countries have large towns where the problems are similar to those of urbanized societies everywhere. It is also important to bear in mind the differences between underdeveloped countries. Some have a much better resource base or a smaller population, such as Argentina, Brazil, Mexico and Malaysia. They are much more likely to be able to overcome their problems than those countries with weak resource base and a large population with rigid traditional ideas and orthodoxy, e.g., Ethiopia, Sudan, Somalia, Niger, Nigeria, Bangladesh and Pakistan.

Population Dilemma of Europe : Although world population has been increasing especially in the developing countries, at an unprecedented rate since the Second World War and there is dismay at the international level about the fast pace of population growth, some of the European countries are facing an opposite domestic concern about population. At present, the salient features of the Europe's population are as follows:

1. It is older than that of any other continent.
2. Population of many of the European countries is stagnating or declining.

3. The fertility rates of the European countries (12) were below the replacement level in 1991 of 2.1 per 1,000. None of the European countries, except Albania, Bosnia and Latvia, is at present is replacing its population through natural increase.
4. The total fertility rates in all the countries of Europe except Albania and Iceland are below 1.5 per cent.
5. West Germany has the oldest population in the world with small proportion of young (15%) and large share of middle-aged (69%) and 16.5 per cent retired people over 65 years of age.
6. Europe's population has begun to decline since 1990.

During the last three decades, the demographers of Europe emphasized on Zero Population Growth—a condition achieved when births plus immigration equals deaths plus emigration. An exact equation of birth and death means an increasing proportion of older citizens, fewer young people, and a rise in the median age of the population.

The zero growth rate of population is creating many problems for the society and economy of Europe. Some of such pertinent problems are:

(i) In some areas, kindergarten, nurseries and schools are closing and universities cut back in the face of permanently reduced demand.

(ii) It is projected that by the end of the century the present unemployment will be replaced by a shortage of workers. Germany, Belgium, Greece, Italy, Spain and Portugal are already facing this problem.

(iii) There will be more burden on the government exchequers as they shall have to provide pensions and social services for about 20 per cent of their citizens older than 65 years of age. Already in Germany there are four pensioners for every ten workers, and it is hoped that by 2025 the number of pensioners will be half to that of workers. If the present pattern of population growth (which is almost zero in most of the European countries) is not reversed, the consequences may be serious and catastrophic, which may endanger the very survival of Europeans.

Schemes for Citizens

The size of population, its structure, composition and growth rate as well as the migration pattern are closely influenced by the population policy of the respective government. Each country has its own population policy. These policies range from encouragement of high fertility to varying degrees of discouragement. Some of the developed countries follow the policy of slow growth, also known as the policy of pronatalism. The countries following the policy of slow growth of population include Canada, the United States, Australia, New Zealand, Japan and Singapore. In these countries, the population growth rate is below one per cent.

This means that their populations are near the replacement level and the birth and death rates are almost equal. In other words, the next generation of children will be about the same size as the parental generation. Contrary to this, the population policies in some of the Afro-Asian and Latin American countries are to reduce the size of population, while in Germany, Russia, Romania, Italy, Austria, France, U.K., etc. where the population growth rate is around or below 0.5 per cent the governments are adopting the policies of population increase. The population policies of the different countries have been formulated to achieve either of the following objectives:

(i) to reduce the population growth rate not necessarily to zero;

(ii) to stabilize the population by achieving a zero population growth rate; and

(iii) to achieve a negative rate of growth with a view to reduce the size of population.

There is a consensus among the scholars of population geography and demography that the only preventive way of achieving any of the above objectives would be to reduce the birth rate instead of permitting the death rate to increase. There is also a consensus that rapid growth of population in the developing countries need to be curbed and for this purpose limiting the births is the best policy.

Policies Discouraging Population Growth : As stated in the preceding paragraphs, the developing countries are mostly following the policies to restrict the growth of population. In fact, during the last four decades, numerous countries have adopted policies aimed at reducing population growth. Economic incentives for small families in the Republic of Korea and the Philippines, financial rewards for sterilization in India, Pakistan, Bangladesh, Nepal and Sri Lanka, the preferential treatment for small family mothers in government hospitals of Singapore, availability of free abortion facilities on demand in Tunisia, Cuba, China, India, Indonesia and Singapore, legalization of abortion on medical grounds in a vast majority of less developed countries are some of the policies which are discouraging for the rapid growth of population.

In most of the developing countries, the techniques for promoting family planning and birth control vary from supplying contraceptives and information in some nations to much more direct involvement in others. Invariably, the promotion involves the communication of new ideas.

It is encouraging to note that an overwhelming majority of the population of the developing countries prefer a decline in the rate of natural increase of population. In most of the less developed countries, the governments have come out with a variety of policies together with mortality control, fertility control, family and individual well-being. It is widely believed that the rate of natural increase can be reduced only if all these measures are combined and the standard of living improves. It is interesting to note that most of the developing countries have fixed the time-bound targets of rate of natural increase by 2010 A.D. For example, Kenya has a target of 2.8 per cent, Mauritius 1.1 per cent, Egypt 1.5 per cent, Tunisia 1.4 per cent, China 0.5 per cent and Sri Lanka 1 per cent.

The Indonesian government, with the help of the United Nations, the World Bank and the World Health Organization, built more than 3,000 clinics in Java and Bali. Mass media such as radio, television, and newspapers, were and still are, used to promote the acceptance of the small family and to draw people to the clinics. In Java and Bali, the first acceptors were older women with many children, but later, as the idea diffused, younger women who wanted to restrict their pregnancies were drawn into the programme. At present, the growth of population in Indonesia is 1.2 per cent per annum.

In some countries, the government's efforts to reduce the birth rate have moved beyond voluntary family planning into various forms of coercion. In Singapore, for example, where space is extremely limited, there have been heavy financial penalties for parents who have had more than two children. Parents have been fined to the equivalent of $100 for the third child, with the size of fine being doubled, tripled, and quadruplet for the fourth, fifth and sixth child. On the other hand, parents who have been sterilized after the birth of the second child have been rewarded with guaranteed education and employment for their children. As a result of this policy, in combination with a rapid economic development, the total fertility rate in Singapore fell from 4.9 in 1961 to 1.4 per cent in 2001. In fact, because the country's fertility rate is well below the replacement level, the Singapore government is now encouraging the better educated women to have more children.

In India, a voluntary family planning policy has been adopted right from the beginning of 1960s, but a tough sterilization policy adopted in 1976-77 under the administration of Indira Gandhi could not bring the desired results. The rigid policy was resisted by the masses. Because India has a democratic form of government and public feelings can be expressed through free elections, objections to the tough policy became evident after it was implemented. Most of the political analysts believe the incumbent party (Congress) lost the 1977 election mainly because of the widespread belief, particularly on the part of Muslims and Catholic minorities, that people were being sterilized against their will.

Subsequently, Indian administration returned to less coercive methods of encouraging family planning. The new methods include placing health experts in all villages with more than 1,000 people, publicizing the merits of small families, and offering financial rewards to parents who accept sterilization after having two children. Many Indian couples, especially those who are educated from the middle income group have planned families. The family planning organizations attempt to provide a variety of contraceptives and other services to them.

During the last four decades, there have been some dramatic decreases in birth rates in different parts of India, but fertility rates differ from region to region, occupation to occupation and one income group to another income group. For the country as a whole, the rate of population increase is still over 1.6 per cent. If this rate continues and China's rate continues to fall, India could become the most populous country in the world by the middle of the 21st century.

The demographic characteristics and the population policy of China are very important to examine as about 21 per cent of the world population lives in this country. Moreover, one of the greatest reductions in fertility rates through governmental policy has occurred in China within the last decade, so it provides an example of the role of government policy in affecting birth rate.

One-Child Policy of China : In the 1970s, the Chinese leaders declared that, in spite of standard Marxist doctrine relating economic power to large labour force, the huge annual increase in population was a major handicap to economic development. The government adopted a more rigid policy and commenced a vigorous programme to reduce family size to two children. By 1980, the goal was changed to only one child per family (except for unusual circumstances and for some minority groups). This goal is being achieved through numerous policies. One policy is to postpone the age of sexual activity. The marriage age is generally over 24 for women and 26 for men, and premarital sexual relations are uncommon.

In China, free contraceptives and abortions are available in clinics throughout the country. From the highest government level down to the smallest rural community, an organizational network exists for implementing the family planning programme. Privacy is limited; women's contraceptive and fertility records are pasted at the local health centres so that any deviation from the norm is noticed. Friends and neighbours may strongly, and repeatedly, urge compliance with the one-child policy.

Social and institutional changes have speeded up the decline in fertility. For instance, better health services reduced infant mortality. More available schooling and the accompanying higher literacy of females were instrumental in reducing birth rates. Also, with the role of women expanding from that of a traditional housewife to include work in a factory or profession, fertility rate declined.

Moreover, the necessity to raise sons for old-age security has diminished. In urban areas, there are retirement pensions for workers. In rural areas, until the early 1980s, the commune guaranteed food, clothing, shelter, medical care and burial.

Under the one-child policy, penalties are assessed against families who have more than one child. These penalties vary with local conditions, but they may include ineligibility for better housing, reduced educational opportunities, delayed food rations, fines, and other economic and social sacrifices.

Despite all these policies and steps, opposition to the one-child campaign does exist, especially in relatively poor rural areas. Dissatisfaction is reflected in the fact that many couples are having more than one child.

Owing to the one-child policy, China is rapidly moving into the fourth stage of the demographic transition. In contrast to the European experience, this shift is being achieved while still remaining largely rural and agrarian. China has accomplished this by creating a social climate where people do not see themselves as independent individuals but instead identify with the state. In effect, the Chinese are practising birth control for the country. A high degree of political organization, as well as social control, economic incentives and public motivation, is a prerequisite for this kind of programme.

The one-child policy of China has its own merits and demerits. It has been appreciated and criticized by some of the leading experts in the field of population studies. The policy is usually justified because of the potentially dire consequences of an increasingly large population. Although the Chinese agriculture produces high yields per unit area and the standard of living and longevity have improved throughout the country in recent decades, there are worries about the future.

Those who support the one-child policy are of the opinion that continued improvement in standard of living can only be achieved by limiting the size of future population. The supporters of the policy also argue that since China is having over one-fifth of the world population, the well-being of future population throughout the world will be affected by the decisions made today about the size of the Chinese families.

The supporters of the policy insist that it is flexible enough to allow exemptions, such as in areas where manual labour is important and in regions of minorities (e.g., the Tibetans, Sinkiang, Inner Mongolia and Uighurs).

Despite all these merits, there are many experts, even within China, who disagree with any policy that will greatly diminish the number of adults in the next generation. According to them, continuous population growth is the key to economic development, which leads to better standard of living, happiness, leisure and increases the life span of the people. In case of low growth of population, the country will suffer from a shortage of workers and military personnel.

The one-child policy has also been criticized because of the possible long-range effects of a 1-2-4 (age structure one child, two parents, and four grandparents). In such an age structure there will be more pampering of child. There are complaints about the excessive amount of pampering that single child receives from parents and doting grandparents. Sociologists speculate that these pampered children, when they become adults, will alter the Chinese society unfavourably. Moreover, a continuation of the policy will mean that in future the higher percentage of elderly people will depend on a smaller portion of working adults. This will lead to high dependency ratio which the pampered child may not bear.

The policy has also been criticized on the ground of individual freedom. Birth statistics by gender in China indicate that female infanticides have occurred in some areas. These statistics seem to reflect a reluctance by some families to abandon the tradition of having several children, especially sons.

An increase in the proportion of elderly persons can affect the future market for various kinds of goods and services, both those sold to young people and those sold to elderly persons. An aging population also enlarges the dependency ratio which, in turn, can affect financial systems, such as social security programme.

Most countries concerned about population growth are attempting to combine economic development with the provision of family planning services. The economic development and social change provide the motivation to have small families. Reducing infant mortality, expanding education systems, broadening the role of women, and achieving economic growth—all have been important catalysts in changing traditional attitudes towards the family. Once the motivation is there, family planning services provide the means for bringing the birth rate down.

Policies Encouraging Population Growth : In general, it is the developed countries in which the population growth is slow and in some regions of France, Germany, Luxemburg, Denmark, Russia, Austria, Finland, Ireland, Italy, Netherlands, Norway, Portugal, Spain and U.K., it is showing a negative growth rate. Usually, the richer the country, the slower the population growth. It has also been established that "development is the best contraceptive".

Looking at the declining growth rate of population, some of the European countries have also started taking positive steps to increase their population. France, for example, grants substantial payments and services to parents to help in child rearing. Eastern Europe is much more active in encouraging growth. Three children per family has become the adopted goal of several Eastern Bloc countries that fear for national survival if present trends continues.

Parental subsidies, housing priorities, and generous maternity leaves at full salaries are among the inducements for multiple child families. Interestingly enough, in Romania, contraceptives are banned and abortions permitted only on doctor's orders in the hope that birth rates can be increased. In general, despite some short-term successes, the Europeans of East and West have ignored both blandishment and restrictions. The consequences will only gradually emerge over the next several decades.

In Soviet Union, in 1950s and 1960s, the birth rate decreased significantly. The low birth rate resulted in the reduced number of youth entering the labour force and the armed forces. This shortage in labour created many socio-economic problems. Consequently, a policy was adopted to encourage large-sized families. In Russia, especially the European Russia, where the rate of natural increase of population in 2001 was -0.2 per cent, the government is trying to create a favourable atmosphere for high birth rate through incentives and publicity.

The Mother Lenin Award was introduced in Soviet Union as early as 1970s. This award carrying a handsome cash, a car and a certificate used to be given to a mother who gave birth to twelve children. In Germany, where the birth rate has fallen as low as the death rate, the government has taken stronger measures to encourage larger families. For example, following the birth of a second child, mothers are allowed one year's leave from work at 80 per cent of their salaries. Low interest loans for homes are available at the time of marriage and when each child is born.

In the U.S.A., where the birth rate is about 1 per cent which is higher to that of most of the countries of Europe, and where a tax allowance is given for each child, there is concern over the implications of a population with a stationary size. Low birth rates combined with low death rates have resulted in an increase in the proportion of elderly people in the population.

At present, about 14 per cent of the U.S. population is over 65 years of age, compared to only 8 per cent in 1950. It is estimated that by the end of 2015, about 15 per cent of U.S. population will be over 65 years of age.

In Canada, a government family planning programme was launched in 1969. Under this policy, prohibitions on the distribution of contraceptives were withdrawn and abortion was made more liberal.

Most of the West European countries have no official population policy, other than the pronatalist policy. In the U.K., it was in 1974 that contraceptives and abortions received official recognition. In most of the Catholic countries, birth control measures have remained illegal, though late marriage, illegal abortions, rhythm methods, etc., have helped these countries in keeping their growth rates low. In France, Belgium and the Netherlands, until recently, dissemination of family planning materials and information was hampered by laws. France legalized contraceptives in 1960s. The birth control devices are still illegal in Ireland, Spain and Portugal. Italy has legalized the pill for medical purposes. It was in 1975 that new laws authorized the government to promote family planning in Italy.

Australia and New Zealand have considered themselves as the under-populated countries. The policies of these governments have been pronatalist and pro-immigration. A policy to have zero growth rate of population has been adopted by these countries.

Japan, Singapore, China and South Korea are the only countries in Asia which have been successful in reducing their birth rates to the level of developed countries, largely by legalizing abortion. The population policy of Japan, adopted in 1950, strongly discouraged a family with more than two children.

In some of the African countries, namely, Congo, Guinea, Ivory Coast, Cameroon, Gabon, Gambia, Mali, Guinea, Togo and Guinea-Bissau, the governments have adopted the policies to increase population. This policy in these developing countries is based on the belief that continued population growth is the key to economic development. In the opinion of the formulators of this policy, a youthful population is necessary to develop the countries of Africa, Asia and Latin America. Some African countries do not advocate birth control because poor living conditions adversely affect both infant mortality and fertility. In many areas, fecundity is low because of poor nutrition or disease and is considered to be a greater problem than high fertility.

Most of the Latin American countries being Catholics in their faith have been very reluctant to accept population growth measures. At present, the Latin American countries are propagating family planning mainly on health and welfare grounds and as a means of reducing illegal abortions. Chile, Columbia, Caribbeans and Central American countries have now adopted clearly anti-nationalist policy. Brazil and Argentina still have policies encouraging high population growth.

About the population policies of China and Indonesia, a detailed account has already been given in the preceding paras. Apart from these two countries, Thailand, Sri Lanka, Hong Kong, Singapore, South Korea, Vietnam, Pakistan, Iran, Nepal and Bangladesh have also well-designed population policies to reduce the birth rates. Countries of the South-west Asia and the newly created Central Asian republics (Kazakhistan, Kirgizistan, Tajakhistan, Turkmenistan and Uzbekistan) are encouraging the population growth. Afghanistan, Iraq, Jordan, Lebanon and Syria have also gone for family planning. Israel continues to favour rapid growth of population. Saudi Arabia has outlawed the import of contraceptives.

Inhabitants of India

India has the second largest population of the world. In March 2001, the total population of India was 1,027,015,47 accounting for about 16.87 per cent of the world population. The growth trend of Indian population during the 20th century has been given in Table.

India: Growth of Population, 1901 to 2025

Year	*Population*	*Total*	*Decade growth Per cent*	*Annual growth rate*	*Growth per cent since 1901*
1901	238,396,327	-	-	-	-
1911	252,093,390	+ 13,697,063	+ 5.75	0.56	+ 5.75
1921	251,321,213	− 772,177	− 0.31	− 0.03	+ 5.42
1931	278,977,238	+ 27,656,025	+ 11.00	1.04	+ 17.02
1941	318,660,580	+ 39,683,342	+ 14.22	1.33	+ 33.67
1951	361,088,090	+ 42,420,485	+ 13.31	1.25	+ 51.47
1961	439,234,771	+ 77,682,873	+ 21.51	1.96	+ 84.25
1971	548,159,652	+ 108,924,881	+ 24.80	2.20	+ 129.94
1981	683,329,097	+ 135,169,445	+ 24.66	2.22	+ 186.64
1991	843,930,861	+ 160,601,764	+ 23.50	2.11	+ 254.00
2001	1,027,015,247	+ 180,627,359	+ 21.34	1.60	+ 331.00
2025	1.330.201.561	+ 352.137.336	+ 22.85	1.50	+ 458.00

Sources: (i) Census of India, 2001

(ii) UNO, 1998-99, World Resources: A Guide to the Global Environment.

Density of Population : The population density of India from 1901 to 2001 has been shown in Table. In 1901, the density of population of India in 1901 was as low as 77 and this steadily increased from 1931 to reach 324 in 2001. The persons living per sq km has increased by 21.3 per cent in 2001 as compared to 1991.

India: Density of Population, 1901 to 2001

Census Year	*Density (per sq km)*	*Census Year*	*Density (per sq km)*
1901	77	1961	142
1911	82	1971	177
1921	81	1981	216
1931	90	1991	267
1941	103	2001	324
1951	117		

Notes: 1. While working out the density of India, Jammu & Kashmir has been excluded as comparable figures of area and population are not available for that state.

2. The density has been worked out on comparable data.

Source: Census of India, 2001, Provisional Population Tables, Paper 1, p. 73.

Growth Pattern : It may be observed from Table that the total population of India in 1901 was about 238 million which rose to 361 million in 1951 and 843 million in 1991. In March 2001, India's population was 1,027 million. The annual growth rate since 1971 has been over 2 per cent, while the growth percentages in 1991 and 2001 over the base year of 1901 were about 254 and 331 per cent respectively.

For the country as a whole, the crude birth rate has declined during the period from 1981 to 1991 from 33.8 to 29.1 per 1,000. Moreover, the crude death rate has declined steadily between 1981 and 1991, from a level of 12.1 in 1981 to 11 in 1987 and 9.7 in 2001. Unfortunately, the infant mortality and maternal mortality rates are high in the Hindi-speaking states of Uttar Pradesh, Bihar, Madhya Pradesh, Rajasthan, Jharkhand, Chhattisgarh, Uttranchal, Himachal Pradesh and Haryana. Poverty, malnutrition, a decline in breast-feeding and lack of sanitation and health facilities are all associated with high infant and female mortality rates.

India's population has multiplied over four times since 1901 and almost tripled since 1941 (in 57 years).

In terms of growth rates, though there has been a slight decline in the rates during the decade 1991-2001 as compared to the previous one, it has been much less than expected in the context of the prolonged family planning campaign and substantial investment of financial and other resources therein. The rate of growth of population in 2001 was for 1.6 per cent.

On the basis of demographic transition the Indian states may be classified into three groups:

(i) Group A, comprising the states of Andhra Pradesh, Goa, Kerala and Tamil Nadu, which have already achieved the replacement level of fertility or are expected to reach that level in the next few years;

(ii) Group B, where the pace of decline in fertility in the past decade has been substantial but not fast enough to make significant dents in growth rates, e.g., Punjab, Haryana, Karnataka, Gujarat and Maharashtra; and

(iii) Group C, consisting of the states of Uttar Pradesh, Bihar, Assam, West Bengal, Orissa, Chhattisgarh, Jharkhand, Uttranchal, Madhya Pradesh, Rajasthan and Jammu & Kashmir where no substantial declines in fertility levels have been recorded in the past decade.

The projected population of the country as estimated by the Planning Commission shows that close to 330 million people are expected to be added in the next 20 years period. This is more than the total added during the previous two decades and almost equal to the population of India at the time of independence.

The uneven growth of population will have a bearing on the political level as the states that have had a higher rate of population growth will have a proportionately large number of representatives in parliament and hence will tend to acquire increasing political leverage. This will be a serious political issue after 2001 when the freeze on the existing number of MPs according to the Act of 1977 will come to an end.

India has 28 states and six union territories. The union territories are relatively small compared to the states, and directly administered by the Central Government. The states vary enormously in size and population. At one end of the spectrum is the mountainous state of Sikkim with a population of only 0.54 million (2001) and at the other end is the state of Uttar Pradesh with a population of 166 million. There are only three other countries in the world, namely, China, America and Indonesia, which have a population that exceeds the population of Uttar Pradesh.

The states vary not only in their geographical size and population count but also in terms of population density, socio-economic conditions, cultural practices, social norms regarding marriage, the status of women in society, and many other factors that influence the health-seeking behaviour and living conditions of the people, especially the productive health of women, and health and survival status of the female child.

The population picture turns kaleidoscopic when the states are looked at individually. During 1991-2001, Kerala, Tamil Nadu and Goa registered growth rates far lower than the national average. These states have also recorded a substantial decline in the growth rate as compared to earlier decades. On the other hand, among the Hindi-speaking states of Bihar, Madhya Pradesh, Rajasthan and Uttar Pradesh, also known as 'BIMARU states, the growth of population continues to be exceptionally high.

Projections until 2025 (with the total population estimated at 1,380 million) reveal that it will be difficult to contain population growth even under assumptions of moderate decline in fertility. It is mainly because of the high fertility of the past and the resultant momentum of population growth. This is especially so in the case of the Hindi-speaking states of the North, even taking into account the fairly rapid future declines in fertility areas in these states as assumed in the projections. The combined population of these states in 2001 has become 463 million.

Sex Ratio : The sex ratio in India is highly skewed, i.e., 933:1,000 (2001). This is largely attributed to women's lower status in Indian society which has contributed to their higher mortality rate in all age groups upto 45. The fluctuating trend of sex ratio may be seen from the fact that in 1901 there were 972 females per 1,000 males which declined to 930 in 1971, 934 in 1981 and 927 in 1991. In 2001 the sex ratio was, however, 933 recording an increase of six females per 1,000 of males.

It is the state of Kerala, where the females have outnumbered males all along, the ratio has been fluctuating. According to the census of 2001, the sex ratio there was 1,058 females per 1,000 of males. The overall deficiency in sex ratio in India can be attributed partly to higher mortality of females and partly to their under enumeration in the census.

Females in India have always suffered from a lower status, right from the time of conception. Women's lower status in Indian society contributes to early marriages, lower literacy, poor nutrition and higher fertility and mortality levels, especially during the reproductive age.

Recently, the large metropolitan cities of Mumbai, Kolkata, Delhi, Chennai and Bangalore have experienced increasing incidence of female foeticide with the use of ultrasonography. The states of Haryana and Punjab are also having high incidence of female foeticide. The sex ratio in Punjab and Haryana has declined from 882 and 865 in 1991 to 874 and 861 in 2001 respectively. It is unfortunate that such modern techniques, meant primarily to identify genetic abnormalities or foetal growth disorders, are grossly misused, in a specific cultural context to perpetuate biases and prejudices against the female child. Concerted efforts at the community and government levels are called for to remove such practices that discriminate against the female child.

Fertility Rate : The total fertility rate in India has gone up. If there had been no contraception, the total fertility rates among married women might now be close to nine children. The increase in natural fertility is mostly due to the relaxation of many traditional checks on fertility that prevailed in Indian society for ages and kept the fertility levels of Indian women well below the biological maximum, or the levels observed in Europe in the 18th and 19th centuries. Such checks are:

(i) prohibition of widow marriage,

(ii) abstinence even among married couples for a considerable period of time in their reproductive life for various religious or cultural reasons,

(iii) prolonged period of lactation,

(iv) terminal abstinence by couples at relatively young ages because of sons and daughters getting married or their becoming grandparents, and

(v) febrile diseases that affect fecundity, such as malaria or tuberculosis, etc.

The relaxation of these traditional checks on fertility, because of modernization and improvements in health of the couples and because of better nutrition and control of certain communicable diseases, have pushed up the natural fertility levels in Indian society. In the absence of family planning programme and increased availability and use of modern methods of contraception, the total fertility rate would have gone up in the Indian society. A more serious and rigorous family planning programme is required to reduce the fertility rate, especially in the rural areas, if the overall growth of population is to be arrested.

Life Expectancy : The average life span of a child born in India has increased over the past four decades from 32.1 years during 1941-51 to 57.3 years in 1981-91 and about 69 years in 2001. This increase is largely attributed to the implementation of various programmes of public health and control of communicable disease after independence.

Among the states, an expectation of life in 1991 of over 65 years has been observed only in Kerala and Punjab. Expectation of life below 60 years has been observed in Assam, Bihar, Gujarat, Himachal Pradesh, Madhya Pradesh, Orissa, Rajasthan, Uttranchal, Jharkhand, Chhattisgarh and Uttar Pradesh. These are also the states where the status of women, especially the female child, has been found to be considerably lower to that of the males.

It is expected that India should achieve by 2005 a life expectancy of birth greater than 70 years and by 2015 greater than 75 years.

Policy for Citizens

The Indian policy-makers realized the importance of population control as early as 1951-52, but a rigid policy was not adopted to arrest the fast growth of population. In 1961-71, the population growth rate was 2.25 which was the highest at any decade after independence. At present, the population growth rate has declined to 1.6 per cent. During the post-independence period, the death rate has been controlled and medical facilities have been extended to the far-flung villages of the country, yet the explosion of population may be attributed to numerous physiological socio-economic and cultural factors.

When the population policy was designed in the First Five-Year Plan of India, it was realized that the base of population is already very large and the trend of population growth cannot be altered easily and quickly. The plan enunciated that the programme for family limitation and population control should:

(a) present an accurate picture of the factors contributing to the rapid increase of population;

(b) discover suitable techniques of family planning and devise methods by which knowledge of these techniques could be widely disseminated; and

(c) give advice on family planning as an integral part of the service of government hospitals and public agencies. The meagre First Plan provision of Rs. 65 lakh for the family planning programme was too little to yield any far-reaching results.

In the Second Five-Year Plan, the voluntary sterilization population policy was introduced. The family planning programme was provided an amount of Rs. 5 crore and it was during this period that 1,650 family planning centres were established in the different parts of the country. Consequently, the family planning programme made an appreciable progress during the Second Plan.

The striking growth rate of population compelled the government to adopt a relatively more clear and less flexible policy of population. It was in this plan that the programme of family planning, involved intensive education, provisions of facilities and advice on the largest scale and widespread popular effort in every rural and urban community. Further, the clinical approach of the first two plans was replaced by an extension education approach aimed at bringing the messages and services to the people in the far off areas of the country through a network of family planning centres.

The masses were educated about the merits of small family and the eligible couples were motivated to adopt the preventive methods of population growth. Moreover, there was more emphasis on education and employment of women. In the Third Five-Year Plan, logistics were provided for family planning which motivated about one million people to accept sterilization.

More emphasis was laid on the family planning programme in the Fourth Five-Year Plan. The most distinctive feature of the Fourth Plan was that it set a time-bound target of reducing the birth rate from 39 per thousand to 23 per 1,000 by 1979. The outlay for the Fourth Plan was raised to Rs. 286 crore. Consequently, by the end of the plan, about 9 million couples were covered under sterilization and about 6 million couples were covered by other family planning methods. About 7 million births were estimated to have been averted during the plan period.

In the Fifth Five-Year Plan, Rs. 500 crore were provided for the family planning programme. The programme sought to integrate most of the basic social services, including education and public health services with family planning and nutrition of children, expectant and nursing mothers. A more rigid policy with an element of compulsion, monetary incentives, penalties and legalization of abortion during the Fifth Plan made the Indian population policy more effective.

High priority was given to the family planning programme in the Sixth Five-Year Plan. The strategy during the plan was to integrate health, family welfare and nutrition services at all levels. Monetary incentives and full rebate in income tax for specified donations for welfare purposes were given by the government. The birth rate was to be reduced to 30 per thousand by the end of 1982-83. The vigorous population policy, followed by the Indira government in the late seventies, was opposed by the masses.

In the Seventh and Eighth Plans, a more pragmatic policy was adopted. There is more emphasis now on persuasion, publicity and family and individual well-being.

Despite all these plans and policies, the population of India is growing at a faster pace and taking the shape of population explosion. The economic development and rising standard of living of some of the people are not adequate to bring down the population growth rate. The time factor is so pressing, and the population growth so formidable, that we have to get out of the vicious circle through a direct assault on this problem. In the middle of the last decade, an attempt has been made to rejuvenate the National Family Welfare Programme. The Ministry of Health and Family Welfare has set-up three market research organizations to conduct independent evaluation of the family planning programme and to make a diagnostic study on the perception, attitudes and practices of the people towards family planning and use of contraceptives.

The revised strategy seeks to broaden the area of family planning by including areas beyond the health sector, such as child survival, women's status and employment, literacy and education and socio-economic development including anti-poverty programmes. It also stresses to make family welfare a multi-disciplinary and integrated effort of all relevant departmental agencies and to make the programme a genuine voluntary people's movement. It is with this objective that the age of marriage is being raised for women from eighteen to twenty years. For raising the status of women, female education is getting adequate emphasis.

Efforts are also being made to involve the voluntary organizations to promote family planning. Committees have been set-up at the state, district, block and panchayat levels to discuss population growth and family welfare projects. None of these steps individually can bring the growth rate low, their package application is essential to achieve the goal of slow growth of population without affecting the declining death rate. The government has to be more serious about the population policy and in case population growth is not checked, all the economic gains through planning will be diluted and India will remain a country of poor and illiterate people.

Citizen's Scheme

The National Population Policy, 2000 (NPP, 2000) affirms the commitment of Government of India towards voluntary and informed choice and consent of citizens while availing of reproductive health care services, and continuation of the target-free approach in administering family services. The NPP, 2000 provides a policy framework of advancing goals and prioritizing strategies during the next decade to meet the reproductive and child health needs of the people of India, and to achieve net replacement levels (Total Fertility Rate) by government, industry and voluntary non-government sector working partnership.

The Purposes

The immediate objective of the NPP, 2000 is to address the unmet needs for contraception, health care infrastructure, and health personnel, and to provide integrated service for basic reproductive and child health care. The medium-term objective is to bring the TFR to replacement levels by 2010, through vigorous implementation of inter sectoral operational strategies. The long-term objective is to achieve a stable population by 2045, at a level consistent with the requirements of sustainable economic growth, social development and environmental protection.

In pursuance of these objectives, the following national socio-demographic goals to be achieved in each case by 2010 are formulated:

1. Address the unmet needs for basic reproductive and child health services, supplies and infrastructure.
2. Make school education up to age 14 free and

compulsory and reduce dropouts at primary and secondary school levels to below 20 per cent for both boys and girls.

3. Reduce infant mortality rate to below 30 per 1000 live births.
4. Reduce maternal mortality rate to below 100 per 100,000 live births.
5. Achieve universal immunization of children against all vaccine preventable diseases.
6. Promote delayed marriage for girls, not earlier than age 18 and preferably after 20 years of age.
7. Achieve 80 per cent institutional deliveries and 100 per cent deliveries by trained persons.
8. Achieve universal access to information/counselling, and services for fertility regulation and contraception with a wide basket of choices.
9. Achieve 100 per cent registration of births, deaths, marriages and pregnancies.
10. Contain the spread of AIDS and promote greater integration between the management of reproductive tract infections (RTIs) and sexually transmitted infections (STIs) and the National Aids Control Organization.
11. Prevent and control communicable diseases.
12. Integrate Indian System of Medicine (ISM) in the provision of reproductive and child health services.
13. Promote vigorously the small family norm to achieve replacement level of TFR.
14. Bring about convergence in implementation of related social sector programmes so that welfare becomes a people-centred programme.

Welfare of Citizens

Although the world population is still growing, it is doing so at a slower rate that demographers had projected only a few years ago. Recent major gains in average life expectancy, reduced rates of child and infant mortality, and the increasing proportion of children now attending school all provide ground for optimism about human well-being. Many of these gains have been made possible by unprecedented rates of economic growth in many countries. Some 3.5 to 4.5 billion people are expected to experience substantial improvements in their standard of living by the end of 2015.

Yet these global success mask urgent, sometimes worsening problems at the local or regional level, especially among developing countries. More than one quarter of the world's population has not shared in the economic and social progress experienced by the majority and still lives in poverty. Hunger, disease, illiteracy, and restricted freedom of choice or action are the persistent problems in many of the least developed countries of sub-Saharan Africa and South Asia, as well as in parts of Central Asia and South America. The pressure of population can contribute to human deprivation, especially in poor rural areas where competition for land and water can strain the capacity of local environments. Rapid population growth is also fuelling problems in many cities, where it can overwhelm the capacity of municipal authorities to provide even elementary services.

Yet the interaction between population growth and human well-being is complex and not a matter of numbers alone. The capacity of countries to support growing population is enhanced when those countries achieve a sufficient, equitable distribution of wealth, technological development, effective government, strong institutions, and social stability.

The world population is still increasing and has now reached over 6 billion which may almost double by 2050.

What are the implications of an approximately doubled population? Population growth is of most concern where countries appear least able to deal with its consequences. Key issues in providing for increased populations will include the achievement of adequate income levels, food security, employment, and the provision of basic social services. Also critical is the sound management of natural resources that, in many developing countries, still directly support the livelihoods of majority of inhabitants.

A significant number of world's people already face critical shortfalls of the essentials needed for a healthy life. Some 1.3 billion live in absolute poverty, 840 million are undernourished, roughly 1.5 billion lack safe drinking water, and about one billion are illiterate. In many areas, population growth is accelerating the rate of degradation of forests, fisheries, and productive soils. Population at direct risk from environmental degradation are concentrated in the least developed countries of the Sub-Sahara Africa, South Asia, and in parts of Latin America. If the future generation in the developed and the developing countries continue to use the resources at the present rate, the consequences in terms of global climatic change, loss of vital renewable resources, and toxic pollution will be severe.

4

Geography and Society

Formation and Structure

The society is made up of elements drawn from diverse origins. Within the population of India are subsumed tribes—no less than three hundred ethnically differentiated communities—caste groups, language and religious groups, displaying striking differences in social organization and cultural patterns, even material cultures. There are differences in racial strands and ethnic and cultural identities are strongly defined. Almost all religions from tribal forms of animism and totemism to Hinduism, Buddhism, Christianity, Islam and Sikhism have their followers in different proportions. The social diversity is perhaps the most powerful manifestation of Indian identity.

The social groups with diverse ethnic origins, representing racial stocks from proto-Australoids and Mongoloids to the different branches of the Mediterraneans, western Brachycephalic groups and the Nordics, found a place for themselves at different points of time adapting themselves to the different ecological niches offered by the physiographic and the climatic setting of the subcontinent.

There are strongly defined tribal identities based on ethnic origins. The tribal population includes within its fold hundreds of large and small ethnic groups from the Santhals, Gonds and Bhils to the Mundas, Oraons, Hos, Todas and Apatanis, to mention only a few. The tribal India is characterized by a heterogeneity of a very high order. Their geographic patterning in the Indian space in itself posed serious problems in the administration of the tribal affairs. The tribes are heavily concentrated in the north-east, in the hills and the plateau surfaces surrounding the Brahmaputra valley as well as in the mid-Indian belt from the Western Ghats to the Rajmahal hills. The tribes have been living in a world of their own in habitats which permitted little interaction within themselves or between them and the rest of the Indian population.

The mutually exclusive domains of the tribal and non-tribal segments of population appear to have contributed to the regional differentiation of the first order. The tribal identity is thus based on ethnicity, as differentiation took place along the ethno-linguai lines. Tribalism implies allegiance to traditional values of social equality, freedom, communitarian ethos, equality of status, including gender equality, segmentation along clan and kinship structures rather than social stratification or hierarchical order governing the status of the different sections of population. The tribal view of social order and social law is essentially different from the rest of the society.

The traditional egalitarian relations, however, underwent a change as the tribal groups were exposed to diverse influences emanating from centres of power, such as the British or Indian administrators, economic classes, such as the moneylender and the petty traders, or the influences exerted by the peasant groups in areas of high population density who slowly spilled over the traditional tribal domains, particularly in mid-India. These exogenetic influences have helped transformation of the traditional tribal relations, with some sections acquiring the traits of an upper class society and other sections losing access to natural resources including land, and thus getting marginalized.

The organization of space in tribal regions may thus be seen as a manifestation of the ways of adaptation to the environmental setting as determined by the historical process of peopling of the traditional habitats by homogeneous clan and kinship groups. This primeval homogeneity originated from their common descent. Over the time, and as a result of their initial occupance of the habitats of their choice clusters of tribal hamlets acquired the character of a monolithic tribal core region. A hill range which defined the watershed, or the channel of a river, served as a boundary separating the core region of one tribal group from the other. This illustrated the process of segmentation. The placement of tribes in spatial segments may be an outcome of a social history of contact, confrontation and contestation between the ethnically differentiated tribes and between them and the peasant groups who hastened the process of their displacement from the river valleys in the course of the colonization of the riverine tracts.

As it would appear tribes have lived away from the river valleys ever since the colonization of the alluvial basins by the agrarian communities. The highlands and the plateau surfaces surrounding the river valleys were not the ideal sites for the pursuit of agriculture. The tribes, although exposed to the agrarian mode, remained on the fringe of sedentary agriculture. The river valleys are occupied by peasant societies whose distinctive trait is a social hierarchical order based on the institution of caste. This hierarchical social order enshrined in the situation of caste is not a characteristic feature of the Hindu peasantry alone. All other sections of the society, whether

Christian or Muslim, are also socially organized along caste lines. Caste is a social phenomenon of great relevance. In fact, caste and communal identity together define the basic parameters of the organization of rural space.

The layout and morphology of the rural settlements reflect the way different caste groups have come together to constitute the village society and have segregated in space in accordance with their function or occupation and their social ranking based on it. The spatial arrangement of caste or community groups within a village in itself is a demonstration of traditional relations based on the single criterion of caste identity.

The propertied classes among these groups emerged as dominant castes, as they were owners of land—the principal means of production. Around the nucleus of landowning castes converged a number of ancillary castes whose position in the village morphological plan reflected their status in the social hierarchy. These ancillary caste groups were subservient to the interests of the dominant castes.

Since the domains of these dominant castes were by and large mutually exclusive, the geographic patterning of the dominant caste groups also revealed the distribution pattern of social authority in rural India. The political ramifications of such a social order cannot be easily visualized.

The dominant castes wielded social power, and when democratic institutions were introduced, monopolized them as they were the ruling elites. This monopoly of democratic institutions and the developmental assets that reached the villages forced the lagging and the deprived sections of the caste society to mobilize themselves politically so as to counter the hegemony of the high castes on the political plane.

The initial process of colonization of the river valleys from the Indus to the Brahmaputra and the Sabarmati and the Godavari to the Kaveri introduced the first level of differentiation in the ethnic content and cultural expression. The river valleys, both major and minor, provided ecological niches for the primeval groups. They might have been derived from the same stem but with distance in time and space, differences in speech form and in culture increased.

The early territories which emerged in the form of *janapadas* around the middle of the first millennium B.C. reflected the onset of this process of social differentiation. Archaeologists, such as Richards and Subbarao, demonstrated how river valleys emerged as regions of attraction, and acquired their own nuances of language and culture. An understanding of the cultural matrix of these so called nuclear regions may hold the key to an appreciation of the roots of cultural diversity that the people of India demonstrated today.

The peasant society based in the river valleys, with all the ramifications of caste based hierarchy, got differentiated adapting itself to the regional ethos. The caste or *varna* system are all-India categories, but their regional expressions differ so much that no two regions are exactly comparable. The ranking of the castes and the perceptions of status within the same caste group (*e.g.*, Brahmans) varies in space, revealing an admirable correspondence with language regions.

Modernising Techniques

Despite the fact that ethnic, tribal and caste identities are strongly defined, the Indian society has been responding to the processes of modernization and social change with varying degrees of intensity. However, the impact of these processes is neither uniform nor all sections of population are equally responsive to change. Urbanization, industrialization, education, social reform movements, political awakening and the advent of democratic processes seem to have exercised an all-embracing modernizing influence. Some Indian sociologists believe that the tradition itself has been modernizing.

Urbanization, for example, is one such process which has brought in its wake a whole series of changes leading to the transformation of traditional way of life, behavioural patterns of the people and to some extent the associated attitudes and value systems. But the impact of urbanization is not uniform. Some regions of India are fast urbanizing while others have remained predominantly rural.

The impact of urbanization is most evident in some two dozen metropolitan centres and a few hundred cities. Even the cities, particularly in the lagging regions of the country, such as Bihar, Rajasthan or Orissa, may reveal the evidence of the continuing hold of rural tradition and the effect of modernization may only be superficial. This is true that in its developed form the urban way of life necessitates changes in the social attitudes, and more importantly, the identity of an individual is likely to be lost. Thus caste and family antecedents are not easily verifiable in a city and the individual may feel relatively free of the social constraints. Moreover, the city intensifies the struggle for survival as opportunities demand stiff competition. The aspiring sections of population acquire new skills through education and training for which cities offer an ideal environment. Paradoxically, the character of Indian urbanization is such that it does not replace the traditional way of life. In fact, it permits the juxtaposition of the modern and the traditional. This is partly because of the continuity of traditional identities based on ethnicity, tribalism, caste and religion which are retained even after migration to the city.

Individuals are distinguishable on the basis of caste, religious and ethnic names which continue to be factors of social reference even in an urban setting, often leading to discriminatory behaviour. In fact, the morphology of towns itself shows the very process of settling down of different social categories in the urban space. The *mohalla, basti* and *tola* formation within the urban centres indicates the division of the urban space between the caste, community and occupational groups. Moreover, the polarization along the communal lines accentuated since the Partition of the country has resulted in the emergence of separate blocks for the two major communities. This is a common feature of the North Indian towns. This is notwithstanding the fact that the impact of urbanization is all too pervasive, leaving its imprints on all spheres of social life.

One notices, on the one hand, that urbanization and industrialization have gone hand in hand generating vigorous rural-to-urban streams of migration and effecting a systematic transformation of traditional occupational structure. Urbanization implies horizontal movement of the rural masses to the urban centres. For example, the migration to the city involves not only a change in residence but also a vertical shift from the primary to the secondary or the tertiary sector of economy. On the other hand, the process of industrialization has been uneven in space and it has left an impact which is spatially fragmented. Industrial complexes have emerged as enclaves revealing little linkage with their rural hinterlands. In fact, their forward linkages are stronger than their backward linkages. But industrialization has also paved the way for a systematic transformation of the rural way of life, particularly in areas of progressive agriculture. The increasing demand for consumer goods in rural areas is an evidence of this ongoing process. In sum, the transformation of agriculture and the ongoing process of rural development have generated a new economic environment. This has led to an increasing degree of rural-urban interaction and a consequent loosening of the hold of traditional values.

But there is yet another side of the story. In the traditional society of India with a definite social structure, when the factory industry came, it conformed to the given occupational structure which was caste-based to a very large extent. Each caste group had a traditional association with an occupation which was hereditary and the labour absorption in industries progressed on the same lines. Several industries attracted labour specific to its needs. The working class was, however, divided along caste lines, trade unionism notwithstanding. Take, for example, the absorption of the Chamars in the leather industry of Agra, Kanpur and Chennai, or the absorption of Dhunas and Julahas (castes traditionally associated with ginning of cotton, spinning and weaving) in the cotton textiles industry. In fact, there were several departments of work in the cotton mill industry, where the untouchables were denied entry because their involvement in work would offend the labourers drawn from the ranks of the middle or high ranking castes.

This implied that the progress in industrialization instead of weakening the traditional caste-based hierarchical structure further re-enforced it. Thus, the impact of industrialization as a modernizing process has not always been positive. In a way the caste system remained intact despite the trade unionism which mobilized workers on the basis of class relations. Class consciousness emerged although it was not sufficient for countering effectively the over-arching influence of the caste-based divisions.

The role of education in social engineering is no less significant. The British introduced a system of education which was essentially geared to their administrative needs in India. Their endeavour was really aimed at producing an educated class of Indians who could be assigned administrative responsibilities at the lower level as lower and upper division clerks and office superintendents. The rate of expansion of British education was painstakingly slow apparently due to a general paucity of jobs in the government sector. The Indian elites drawn from the feudal and the princely houses, however, initiated their children, mostly males, into the British type of education. The scions of the Indian princely houses preferred to join the prestigious institutions in England more as a status symbol than out of any economic compulsion.

This new type of education, even after the establishment of the presidency colleges at the port centres and the degree colleges at the district centres elsewhere, did not have much of an appeal to the Indian people. The masses remained outside the ambit of modern education. Even the general literacy rates remained low despite the expansion of education at various levels from the primary to the graduation levels.

The situation did not change even after independence and despite repeated proclamations to universalize education, it remained by and large confined to the upper echelons of society. In fact, there is no gainsaying the fact that the demand for education is intrinsically linked to the returns from education. Peasantry and the working classes which constitute an overwhelmingly large chunk of population did not see any value in the kind of education imparted to them ever since the days of the British Raj. Moreover, rampant unemployment further disillusioned the first-generation learners and detracted them from seeking education as jobs became scarce and competition for jobs intensified. It may be noted that today the character of higher education is basically elitist. Moreover, it has a strong urban orientation. The result is that the benefits of education have mostly accrued to the elites only comprising the upper classes and the urbanized high castes.

The expansion of education in India since independence has been widespread, although its impact has not been uniform. In fact, education has not been able to play its constructive role as a catalyst of social change. Education has also failed to weed out the traditional values based on superstition and obscurantist philosophies.

Theory and Views

The term 'social geography' carries with it an inherent confusion. In the popular perception the distinction between social and cultural geography is not very clear. The idea which has gained popularity with the geographers is that social geography is an analysis of social phenomena as expressed in space. However, the term 'social phenomena' is in itself nebulous and might be interpreted in a variety of ways keeping in view the specific context of the societies at different stages of social evolution in the occidental and the oriental worlds. The term 'social phenomena' encompasses the whole framework of human interaction with environment, leading to the articulation of social space by diverse human groups in different ways. The end-product of human activity may be perceived in the spatial patterns manifest in the personality of regions; each pattern acquiring its form under the over-arching influence of social structure. Besides the patterns, the way the social phenomena express themselves in space may become a cause of concern as well. This has attracted scholarly attention, particularly since 1945 when all-embracing changes in the political and economic order of the world started casting their shadows on the global society. As compared to the other branches of geography social geography has a certain amount of recency. Eyles saw the antecedents of contemporary social geography in the development of the philosophy of possibilism in the late nineteenth century. The view of social phenomena is all-embracing and holistic, based on the totality of human interaction with environment. Eyles also visualized social geography as a continuation of the philosophy of Vidal de la Blache and Bobek:

> "...it stressed both the humanistic nature of the geographical world... and the classificatory nature of... human geographical work...".

Upto 1945, social geography was mainly concerned with the identification of different regions, themselves reflecting geographic patterns of association of social phenomena. In fact, during the twenties and the thirties of the twentieth century, social geography started its agenda of research with the study of population as organized in settlements, particularly urban settlements. This was understandable as population in the Anglo-Saxon and American world was overwhelmingly concentrated in the urban areas. The process of urbanization had thrown up issues of social concern such as access to civic amenities and housing and the related socio-pathological issues, such as incidence of crime, juvenile delinquency and other expressions of mental ill-health. Socio-geographical studies of population distribution and ethnic composition in urban areas emerged as a major trend during this phase.

The underlying idea was to examine the social content of the urban space which resulted from coming together of diverse ethnic groups within a city. The city with its specific functional specialization cast these social groups in its mould, resulting in the assimilation of diverse elements into a universal (Europeanized) urban ethos. However, certain ethno-cultural identities (e.g., Blacks in American cities, North-Africans in France and Asians in Britain) were so strongly defined that they continued to defy the forces of assimilation. Moreover, the land market in the Anglo-Saxon and American cities further marginalized these coloured people. This resulted in their spatial segregation in ghettos with all the socio-pathological implications that follow from it. Emphasis on population characteristics remained a major preoccupation of social geographers till the fifties of this century. During the fifties, the tradition continued with social geographers mainly preoccupied with population characteristics. Social geographers differentiated between regions on the basis of the dominant patterns as social phenomena, mostly based on the population characteristics. Later, under the influence of the rising tide of quantification, social geographers started employing area-specific data in order to discover spatial patterns.

During this phase of development, the major focus of research remained on the analysis of the social data for the cities. Social area analysis emerged as the main tool of analysis.

One inevitable consequence was that studies in this area, such as factorial ecology, made social geographic research dependent on the theories of human ecology. As against this Emrys Jones' study of Belfast gave due consideration to the role of values, meanings and sentiments in locational activity. It may, however, be pointed out that any study of the social phenomena within the city in the context of factor analysis helped only in the identification of patterns.

The Definition : The taxonomy of a discipline, while arising out of its logical system, subsumes within itself the specificities of its intellectual tradition, whereby words and terms acquire specific connotations and nuances of meaning through large-scale usage and social acceptance. But this process of crystallization of the classificatory scheme is greatly distorted if the same term tends to acquire different connotations or different shades of meaning tend to be expressed through the same term.

Such is unfortunately the case with that segment of geographical studies which is termed as Human or Anthropo or Social or Cultural Geography. The term "Human Geography" has a vintage value; it emerged in an embryonic state as an element in the essential dichotomy of geography during the classical period itself and acquired more definitive connotation at the hands of the great French possibilists. The term "Anthropo-Geography", on the other hand, arose within the rigid and inflexible conceptual framework of environmental determinism.

The term "Social Geography" was perhaps introduced by Vallaux in 1908 through his *Geographie Sociale: La Mer* as a synonym for Human Geography and has since then remained ill-defined—its boundaries fluctuating at an alarming rate. The term "Cultural Geography" is a gift from the new world, which, while contributing a new item in geographical glossaries, has unfortunately only added to the semantic confusion. A look at some of the standard definitions of these terms would clearly bring out the prevailing lack of clarity on these questions.

Monkhouse in his *A Dictionary of Geography* defines Human Geography as the "part of Geography dealing with man and human activities". In the same volume, the learned scholar states later that Social Geography "is often used simply as the equivalent of Human Geography, or in the U.S.A. as 'Cultural Geography', but usually it implies studies of population, urban and rural settlements, and social activities as distinct from political and economic ones". Dudley Stamp in the *Longman's Dictionary of Geography* defines Cultural Geography as "that which emphasizes human cultures and is commonly equated with human geography".

> It is quite clear that definitions like the above are of no help in demarcating the areas covered by these sub-disciplines of Geography. If there is so much of connotative similarity there is a strong case for discarding two of these terms so that geographers can at least understand each other.... Alternatively, two of these areas of academic work may be viewed as sub-sets of the third one.
>
> —Moonis Raza, *A Survey of Research in Geography* 1969-72, Bombay: Allied, 1979: 63-64

It is a noteworthy fact that the western social science was alive to the real issues in society. Social geography also could not remain unaffected by these trends. Thus, social geography in the western world developed much in response to political happenings of contemporary social relevance. For example, social change in attitudes and the perception of prevailing reality necessitated corresponding changes in the theoretical framework adopted by social scientists. The American society, for example, was overwhelmingly influenced by the war in Vietnam. A common concern was expressed on issues such as poverty and social inequality within the United States. The social relevance movement in the contemporary social sciences also affected geography and issues such as race, crime, health and poverty received an increasingly large attention.

Emrys Jones and John Eyles who described social geography as a group approach conceded that the attempts at definition represented the viewpoints of their authors to which others may not agree. The progress of social geography in the decades since 1960 has taken three main paths, each cluster of research acquiring the status of a school of thought in its own way.

(a) A welfare or humanistic school mainly concerned with the state of social well-being as expressed by territorial indicators of housing, health and social pathology largely within the theoretical framework of welfare economics.

With due apologies to those omitted by oversight, the last twenty-five years or so have produced eight definitions of social geography, seven of which are provided by geographers working in the Anglo-American tradition. These are: the identification of different regions of the earth's surface according to associations of social phenomena related to the total environment (Watson, 1957: 482) the study of the patterns and processes (required) in understanding socially defined populations in a spatial setting (Pahl, 1965: 81) the study of the areal (spatial) patterns and functional relations of social groups in the context of their social environment; the internal structure and external relations of the

> nodes of social activity, and the articulation of various channels of social communication (Buttimer, 1968: 144) the analysis of the social patterns and processes arising from the distribution of, and access to, scarce resources and... an examination of the societal causes of, and suggested solutions to, social and environmental problems (Eyles, 1974: 65) the understanding of 'the patterns which arise from' the use social groups make of space as they see it, and of the processes involved in making and changing such patterns (Jones, 1975: 7) (it) stressed structure relations in the analysis of social problems...
>
> Analysis (is) based on interrelated material reality and the social contradictions this produces; which are seen as the motive force for change, and thus responsible for the development of problems like different level-of-living conditions (Asheim, 1979: 8) the study of consumption, whether by individuals or by groups (Johnston, 1981: 205) it is an interactionist perspective which aims to uncover how social structure is defined and maintained through social interaction, and which studies how social life is constituted geographically through the spatial structure of social relations (Jackson and Smith, 1984: vii)
>
> —John Eyles, *Social Geography in International Perspective,* London: Croom Helm, 1988: 4-5

(b) A radical school which employed Marxian theory to explain the basic causes of poverty and social inequality. This school of thought related the contemporary social problems to the development of capitalism particularly the internal contradictions of capitalism. For example, cities and the communities within the city were perceived as organized spatially in response to the class relations and the Marxian interpretation was that a welfare approach might not be helpful.

(c) A phenomenological school which laid an extraordinary emphasis on *lived experience* and on the perception of space by social categories based on ethnicity, race or religion.

It is thus obvious that contemporary social geography is in line with the theoretical development in human geography as a whole. This does not mean that the welfare or humanistic concerns or the quest for the causes of social inequality and class-based exploitation or phenomenological perceptions of space have replaced the tradition of areal differentiation or region formation. All these approaches have continued to co-exist.

Some themes have received greater attention at certain stages in the development of Anglo-American school of geography. They may be mentioned here in brief.

References have been made to *social physics*, implying that for the analysis of human behaviour analogies can be drawn with the physical world. Around the middle of the nineteenth century Auguste Comte adopted an approach assuming that principles of physics, or mechanics, may be applicable to human society as well. The idea was revived by J.Q. Stewart in the forties of the twentieth century. In association with William Warntz the two developed the theory of social physics to create the field of 'macro geography'. Based on these concepts a gravity model was developed in human geography which tried to explain interaction between places illustrated by, for example, the movement of people and goods as products of the mass (population size etc.). A distance factor also operated such as cost as an exponent revealing an inverse relationship.

The basic idea of the gravity model also found its place in other models such as entropy maximizing model and diffusion model. Other applications are found in rank-size rule and the population potential model.

These approaches were contested by many geographers who found social physics as simply mechanistic. The human society was not exactly a physical organism which could correspond to precisely defined laws.

The study of social phenomena, spatially variegated as they were, led to the identification of social areas and a *social area analysis* followed. American sociology adopted social area analysis as a technique for relating social structure with urban patterns. In this connection reference may be made to the pioneering work of two American sociologists, Eshref Shevky and Wendell Bell. The two hypothesized that within a city the range and intensity of relations depends on the social rank; that the process of urbanization leads to differentiation in the functions of the households leading to changes in the family status; and that social organization within the city leads to concentration of groups along cultural and ethnic lines. Thus, the ethnic status of an individual also plays a role in social interaction. Geographers who adopted social area analysis as a method in their studies of urban social geography depended on statistics disaggregated for the micro units such as the census tracts within the city. Variables were chosen to represent the three constructs of social rank, urbanization, and segregation in order to develop a composite index on the basis of which census tracts could be classified.

The technique was criticized for being mechanistic as there was no link between the social scaling and differentiation of population within the urban space. It was argued that the three constructs themselves were inadequate to portray the urban social reality. As a method social area analysis was abandoned in favour of what came to be known as factorial ecology. Its importance, however, lies in the fact that at a certain stage in the historical development of social geography it played a highly seminal role furnishing a basis for systematic analysis of urban social space.

Western social geography, particularly the school of thought pursuing social welfare approach, attached the highest importance to the concept of social well-being. It was hypothesized that well-being characterizes a state in which the basic human needs of a given population are satisfied because the people have sufficient income for their basic needs. However, the concept was defined within the framework of the social system of capitalism. The high income groups organized themselves in space in such a way that their basic needs were optimally satisfied. Income played a crucial role in creating optimal conditions for the social infrastructure, such as housing, civic amenities, health, education and recreation. The underlying assumption is that the poor were not in a position to satisfy their basic needs. The state of well-being is achieved only when the income is sufficient to fulfil the basic needs, meaning thereby that the poverty has been eradicated and when the services are available to all sections of the society on a sustainable basis.

It may be noted that both the western social science and social geography were alive to the real issues in society and the social scientists, including geographers, responded to political happenings and the social implications of these happenings attracted their attention. While the Indian social sciences, particularly sociology, social anthropology, political science, economics, education, social linguistics and contemporary history, have been alive to the emergent issues in the wake of political, social and economic development since independence in 1947, geographers in general and social geographers in particular have not evinced much of an interest in the contemporary issues of national interest. The first generation of Indian geographers, *viz.*, George Kuriyan, S.P.

Chatterji, S.M. Ali, C.D. Deshpande followed by V.S. Gananathan, and V.L.S. Prakasa Rao debated extensively the issues of national reconstruction, suggesting planning strategies for the optimal development of the nation and the regions by better and more efficient utilization of natural resources.

However, their debates mostly remained internal to geography, although echoes were heard in the corridors of power, *e.g.*, Planning Commission. Exceptions apart, no significant purpose was served by these debates as a meaningful dialogue could not be conducted on a sustainable basis between geography and other social science disciplines. This thwarted the process of cross-fertilization of ideas across the disciplines, with the result that social geographical research suffered a major setback. Geography was not only marginalized, all possibilities enabling it to make a contribution to the critical social theory were also denied to it. Ensonced within the confines of its own academic shell, it was virtually reduced to a social isolate.

In the seventies of the twentieth century, the Centre for the Study of Regional Development at the JNU emerged as a new nucleus of research with a vast potential for a dialogue with other social sciences. Issues such as tribal under development, agony of the masses hit by a syndrome of droughts, scarcities and famines, poverty, particularly rural poverty, social under development as expressed in illiteracy and levels of educational backwardness, de-stability in tribal areas in the wake of developmental projects, displacement of people by big river valley projects, disparities in levels of development in drought-prone, mountain and hill areas, etc., received increased research attention.

This new academic environment enriched geography's adaptability to social science discourse. In a way the JNU experiment laid down a new agenda for social geographical research building systematically on the tradition of V.L.S.

Prakasa Rao and his associates who focused their efforts on the problems of perspective planning for national and regional development. Social geography at the JNU paved the ground for more give-and-take between disciplines, enabling geography to find a place in the realm of Indian social science.

Social Sciences Relationship

Social geography has genetic relations with other social sciences, particularly with social anthropology, sociology, social history, archaeology and socio-linguistics. Given the fact that social geography has been marked by a late arrival on the Indian academic scene, not much give-and-take has taken place across the disciplines. Unlike the western scholars, Indian social geographers were not greatly attracted to the sociological theories, nor the post-colonial, post-modern discourse influenced them much. Thus, they remained aloof to the major developments in critical social theory.

There is no gainsaying the fact that social geography in its evaluation of social structure, social processes and social transformation cannot flourish without productive and reciprocal interaction with sister social sciences as mentioned above. Take, for example, the case of *archaeology.* The evolution of perennial nuclear regions, and the role of the river valleys in supporting agrarian communities from Neolithic to Chalcolithic and the eventual rise of *janapadas* in the same river valleys received extensive scholarly attention in several disciplines dealing with Indian prehistory. Archaeology furnished the basis of analysis, correlating cultural development with environment and the routes of internal migration. Archaeologists such as F.J. Richards, B. Subbarao, Mortimer Wheeler and H.D. Sankalia and historians, such as D.D. Kosambi and K.M. Panikkar, evaluated the vast evidence in the form of human artefacts (tools, implements and other records of material cultures) to explore the link between cultural development and environment.

Geographical accessibility, or lack of it, played a critical role in the diffusion of human cultures. Archaeologists in their turn were immensely benefited by developments in the geological science (stratigraphy, palaeontology and methods of carbon dating) as well as in palaeo-climatology. This vast fund of evidence was available to the geographers in their reconstruction of the past geographies. Of particular interest is Spate's interpretation of cultural regions as evidenced in the history of human occupance of land. S.M. Ali's reconstruction of the geography of the *janapadas* is another example of this interdisciplinary research. This treatise in social geography draws insights from these contributions to the prehistoric scene in India as interpreted by eminent archaeologists, historians and historical geographers.

Just as physical geography derives its basic source materials from other physical sciences, such as geology, meteorology, pedology and biology, social geography shares a common ground with *social anthropology* and *sociology*. As a field of study anthropology is concerned with the dimension of human relationship with Nature. Social geography with its emphasis on spatial variation in human interaction with Nature naturally stands face to face with anthropology. However, methodologies differ strikingly. While the latter accumulates systematic knowledge and space is only incidental to its basic logic, the former sustains itself through exploring and developing the evidence on the spatial correlates of social phenomena.

While in the early phase the knowledge of the Indian social structure and its constituents, such as tribes, castes, rural institutions, folk customs, folk religion, social organization and material cultures, developed through ethnographic studies, anthropological research acquired new dimensions after independence. The tradition of British (and European) ethnography in India was already more than a hundred years old before the Ethnographic Survey of India was established in 1901.

As a result of this initiative, Indian tribes and castes, which had already been extensively studied, received an increased attention. A series of ethnographic studies on the people of India appeared in the provincial reports on North-Western Frontier Province, Punjab, United Provinces of Agra and Oudh, Central Provinces, Bengal and Madras Presidencies and other parts of South India. These materials formed a substantial basis for cross-country comparisons in a social geographical context. Anthropologists focused their attention on questions of race and caste, cultural change among tribes, kinship, as well as on village studies.

The Ethnographic Survey of India, re-christened as Anthropological Survey of India in 1946, extended this work further. Anthropologists working within the thrust areas of the Anthropological Survey of India generated valuable data on tribes and castes in the changed context of India's independence and the developmental opportunities thrown up by this political change. Foreign students of Indian society, particularly American and French scholars, also contributed to this research by analyzing the institutional set-up of the rural society, tracing the impact of modern technology on these institutions and the changes coming in the wake of the community development programmes launched after independence.

Studies conducted by the Anthropological Survey of India provided a vast fund of literature not only on the Indian tribes and other ethnic groups, their social characteristics and cultural change as a result of their interaction with non-tribes, but also for meaningful social geographical interpretation. Anthropologists also studied Indian material culture in its variegated forms—settlement types, housing patterns, tools and implements, rural arts and crafts, production techniques, folk arts and dances, clothing and ornaments, and the like. The massive work done under the People of India Project led by K.N. Singh is an evidence of this continuing interest.

It may be noted that social geography emerged so late on the Indian scene, and its development was so retarded that much of this vast fund of literature remained unutilized, brilliant insights provided by O.H.K. Spate in his *India and Pakistan* notwithstanding. A glorious opportunity to develop models of inter-regional variation and regional integration was missed. Moonis Raza in collaboration with this author handled massive data on Indian tribes, drawn particularly from 1961 and 1971 census returns to highlight the salient features of the tribal reality. In their *Atlas of Tribal India*, which was much more than an atlas, they presented the tribal problem in a spatial/regional context. However, their effort did not make much of an impact on other sister disciplines in India . Another treaties in social geography, *viz.*, A.B. Mukerji's *Chamars of Uttar Pradesh* was also entirely based on the census data.

Contributions to Indian sociology in the initial phase after independence came mainly in the area of village studies. Studies of village communities and their institutions, such as caste and religion, became the main thrust of research. Caste received an increased attention because of its role in the Indian society, particularly in the changed context. Its origin and historical development attracted the attention of many sociologists.

It may, however, be pointed out that as far as India was concerned there was no clear distinction between sociology and social anthropology. Many sociologists believed that the two disciplines were so close to each other that there was hardly any difference. M.N. Srinivas believed that the nature of Indian society was such that no clear distinction could be made between sociology and social anthropology.

In the words of M.S.A. Rao, "in the Indian context a sociologist can ill-afford to make any artificial distinction between the tribal and folk society and the advanced sections of the population, that is, between social anthropology and sociology, nor can he (she) confine himself (herself) to any single set of techniques. Trends of research in sociology and social anthropology in the post-independence period favoured close association between them".

The pre-independence history of development of social sciences in India also shows a close association between sociology and geography. An eminent British geographer and town planner, Patrick Geddes, was called upon to establish a department of sociology at the University of Bombay in 1919. His writings on India bridged in a big way the gap between geography and other social sciences. As a result, the sociological research also focused on urban studies and town planning. A perusal of later studies shows that this glorious tradition of interdisciplinary research did not lost long. This was, however, a disturbing trend.

In the post-independence period Indian sociologists and social anthropologists also focused on community development programmes, elections, communalism, *dalit* movements and the role of rural leadership. This branch of sociology, which came to be known as political sociology, bridged the gap between political science and sociology. While interests in rural studies, caste and urban studies continued, sociology diversified its area of inquiry by incorporating problems of Indian civilization, such as social and cultural change, which were often described as Sanskritization and Westernization.

It may be pointed out that these areas were of critical interest to a social geographical interpretation of India. These researches helped immensely in the development of socio-geographical insights and paved the way for mutual give-and-take between geography and sociology. Evidently, the underdeveloped state of social geography thwarted such a process. There was no evidence of such a give-and-take. However, geography could find a place in the Surveys of Research in Social Sciences initiated by the Indian Council of Social Science Research. Academic interaction between disciplines such as social geography, social and cultural anthropology and sociology is also evident in researches conducted at the Centre for the Study of Regional Development, of the Jawaharlal Nehru University and at the departments of geography at several universities, such as Punjab, Pune, Mysore and BHU.

Socio-linguists often talk of ecology of language and use ecological criteria in their attempt to understand affiliation of the people of India to a large number of spoken languages and dialects as evident in space in different parts of the country. Their primary interest is in the 'everyday life speech activity' as expressed in communication and institutional networks. They are also concerned with the plurality of cultures and languages evident in the South Asian region. The linguists, more particularly socio-linguists, collect data on the variety of spoken languages, classifying them into groups, sub-groups, branches, sub-branches and families. Their classificatory schemes often lead to an understanding of the geographic patterning of the speech communities in India. Their studies also point out the changes which have taken place in the course of time in the geographical location of the speakers of different languages, indicating either displacement or assimilation.

These changes often reveal the pattern of language maintenance or shift as well as bilingualism or multilingualism as a consequence of inter-communication with other groups. The linguistic plurality owes its origin to the processes, such as migration, resulting in contact and communication, as well as to the super-imposition of languages through administrative mechanisms. It is thus evident that research in socio-linguistics is of direct relevance to the social geographers interested in the problem of language distribution. These studies furnish evidence of the origins of India's linguistic plurality and the communication networks of the contemporary society. In fact, contributions to linguistics and socio-linguistics have helped the development of the field of linguistic geography, sometimes referred to as geo-linguistics. Spatial aspects of language distribution have attracted the attention of geographers, including the Indian geographers. Research completed at the JNU may be mentioned as an example.

Linguists and socio-linguists, such as Suniti Kumar Chatterji, S.M. Katre, B.L. Sakharov, Murray B. Emaneau, L.M. Khubchandani, D.P. Pattanayak, Colin P. Masica, S.N. Mazumdar, Anvita Abbi and E. Annamalai have made valuable contributions to our understanding of the language scene in India. Reference may also be made to the pioneering work of G.A. Grierson, particularly his *Linguistic Survey of India*. The linguists attached to the 1961 Census of India also deserve special mention. They organized the multitude of linguistic data collected during the 1961 census operations. These contributions helped in understanding many crucial issues related to the language identity of the people of India. Notable among them were problems of language classification, language maintenance and language shift. Other related issues, such as geographic patterning of bilingualism and multilingualism, communication in the changing socio-political environment and linguistic plurality were also discussed extensively by the linguists and the socio-linguists.

As the above survey shows the field of social geography is inadequately developed in India. In fact, it is taking its roots and is likely to flourish as a major thrust area in the coming decades. Its scope and purpose will be discussed sub-sequantly which presents the themes of social geography in an historical perspective, looking into the present through the mirror of the past.

5

Geography and Religion

India presents a baffling diversity in religious persuasions and faiths. Although the traditional religion of the land is Hinduism, many other faiths and belief systems, from tribal forms of religion to Buddhism, Christianity and Islam, have coexisted for centuries. They have created for themselves cultural niches within a shared space. These faiths are indigenous (Indic) as well as introduced from outside (extra-Indic). The Indic religions have all evolved from early Hinduism, which has been undergoing changes in content and ritual practices in response to the prevailing cultural, ethno-lingual and ecological diversities in different regions of the country. Within Hinduism, a number of sects, such as Vaisnavism and Saivism, emerged on the scene adding further diversity to the cultural mosaic. These sects have a specific geographic patterning of their own in the country. This shows how ideological differences and philosophical interpretations lead to diverse socio-cultural practices and the associated rituals based on religious faiths.

Thus it is evident that the differences in religious ideologies may lead to sect formation even within the same religion. These differences have led to the emergence of regional nuances in religious practices. In the same way, protest movements within Hinduism eventually led to the emergence of new faiths, such as Buddhism and Jainism. These protest movements also led to reform within Hinduism. Another Indic religion—Sikhism—was genetically linked to these changes. In its original form it was a fine blend of the basic elements of Hinduism and Islam. Religious faiths which originated in West Asia, *e.g.*, Christianity, Zoarostrianism and Islam, also found their followers in India. Initially following a sea route to India both Christianity and Islam had their early base in the littoral regions of South India, particularly the western coastal region, from where they spread into the interior parts. Later missionary activity systematically organized by the Christian missions during the subsequent centuries, particularly the eighteenth and the nineteenth centuries, found many responsive groups among the tribes and the Hindu depressed castes.

The spread of Islam in northern and western India was a later phenomenon. In fact, it followed the sequence of developments leading to the Muslim conquest of the northern parts of the country in the medieval period. Unlike the Christian missionary activity Islamic enterprise was never systematically organized.. However, the emergence of Muslim seats of power in the different regions of the country provided an added incentive to conversion. Evidently, coexistence of multiple faiths and a pervasive spirit of tolerance have been a distinguishing feature of the Indian society through the ages.

Presentation of Religious Aspects

Although religion is a matter of personal faith, religious identity of an individual in India is often expressed at the social plane. Unlike the western world where mass celebration of the religious occasions is infrequent, more so the public display of festivities is rare, religious celebration in the oriental societies is a sociological phenomenon. The Indian society conforms to this norm. Mass festivals are common and the year is dotted with religious events, major as well as minor, which have far-reaching social implications. In fact, in some religious groups, there are several occasions during the year when people publicly display their adherence to a certain form of religious ritual. For example, large processions are taken out with great zeal. Likewise, people offer mass prayers in mosques and churches on certain festivals. Mass prayers are also held in mosques on every Friday. The celebration of festivals acquires dimensions which transcends the limits of the private sphere of life. They become occasions of public expression of religious identity. These practices have continued and coexisted giving strength to the pluralistic nature of India's religio-cultural ethos.

Elements of Religious Identity : The issue of religion, or religion-based identity, may be approached in several ways. First, religion is a matter of faith, a personal affair and a philosophy of life of an individual or a group of individuals. Secondly, and flowing from the first, followers of a certain religion, by virtue of their adherence to a common faith, may develop a community feeling. This leads to conscious or unconscious expression of solidarity with the followers of that religious faith. This community feeling characterizes all religious group formations. Thirdly, a common code of social conduct based on a religious faith may lead to public expressions of a particular religious identity, *e.g.*, dressing in a particular way, avoidance of certain items of food and mass assemblies and public demonstrations on religious occasions. These divergent codes of conduct and socio-cultural practices eventually lead to a consciousness of religious and cultural differences. Followers of other religions are often ranked on a scale constructed in the light of a particular religious ideology and are then rated as superior, inferior or even untouchable. If religion is a matter of personal faith, and there is little adherence to a publicly manifested code of conduct, it does not affect anybody. But if there is a public assertion of one's religious identity expressed pronouncedly in dressways, foodways, avoidance of or preference for certain items of food, cooking habits, eating habits, inhibitions in inter-dining and so on, a religious group acquires an identity of its own and gets differentiated from other religious groups on a social basis. Other forms of discriminatory social behaviour follow.

This promotes, on the one hand, an internal feeling of solidarity and, on the other, a feeling of division. Such divisive tendencies may acquire acute forms and may result in a variety of social conflicts. When groups are formed and perceived on religious basis social authority is often likely to be wielded by a priestly class (the authority of the church as in Roman Catholics or the authority of the *Ulama* as in Islam, although the two are not comparable in orientation and rigour).

This sooner or later acquires political nuances. When such a stage is reached, religion is likely to become the basis of social mobilization which eventually may lead to a social discord. The highest stage in this regimentation is reached when a state identifies itself with a particular religion and subjects the followers of other religions, or religious minorities, if any, to a discriminatory treatment. Theocratic states based on dissimilar religious formations may eventually confront each other in situations of war in an attempt to subjugate each other. Crusades were the best example of such conflicts when the contesting parties were fighting in the name of religion.

The above discussion shows that religion in the world societies has not always been purely a matter of simple faith. Actions of individuals or groups often transcend the limits of personal space. Then religion becomes a manifested basis of social differentiation. In that role, it has far-reaching operational implications which do not always seem to hint at harmony or cohesion. At the same time, there is no gainsaying the fact that in the world society, religion as a moral philosophy has played a role of promoting harmony, peace and commitment to civilized public behaviour.

The values inculcated by religious teachings are universal and reveal the essential unity of all religions. Religion has acted as a civilizing force promoting humanism, respect for other forms of identity and a spirit of sacrifice in an endeavour to achieve higher goals for human coexistence. Religion has induced individuals to subordinate themselves to the higher ideals of humanism and sacrifice their own comforts for the collective good of the humankind. However, human history is also full of instances when religion-based spirit of harmony and compassion has been largely ignored in order to establish the supremacy of a given religious formation.

The same factors which are cohesive in a given formation become the basis of rivalry and competition between the different religious groups. While it is true that religion promotes spiritualism, spiritual values have often been superseded by the human lust for secular authority and material well-being. Despite the religious teachings for good moral behaviour conflict between the good and the evil has continued through history.

Historical Background

Let us now examine the place of religion in Indian society in its historical context. Religion is a form of social organization. While it is personalized, social expression of one's religious identity leads to significant behavioural patterns. The attitudes adopted by different religious groups reveal that ideology is a determining force in social behaviour. People chalk out their social interaction modes in the light of the religious faith they profess. The ramifications of a differential pattern of social behaviour are seen in patterns of social interaction, celebration of festivals, organization of cultural activities, management of social space and personal manners. Western education has brought about a certain degree of social transformation, nonetheless the hold of religious ideologies is too strong. At the same time, it is true that an enlightened class of Indians is thoroughly secular and capable of rising above the religious identities in all situations of crisis.

> ***Mother Earth :*** That the worship of chthonic deities such as 'Mother Earth' generally presupposes a relative importance of agriculture is fairly obvious, but such parallelism is not always direct. Nor can it be maintained that the heavenly gods, as representative of a hero's paradise beyond the earth, have everywhere been noble gods rather than chthonic deities of the peasantry. Even less can it be maintained that the development of 'Mother Earth' as a goddess parallels the development of matriarchal clan organization. Nevertheless, the chthonic deities who controlled the harvest have customarily borne a more local and popular character than the other gods who reside in the clouds or on the mountains is frequently determined by the development of a feudal culture, and there is a tendency to permit originally tellurian deities to take their place among the gods who are resident in the skies. Conversely, the chthonic deities frequently combine two functions in primarily agrarian cultures: they control the harvest, thus granting wealth, and they are also the masters of the dead who have been laid to rest in the earth. This explains why frequently... these two most important practical interests, namely, earthly riches and fate in the hereafter, depend upon them. On the other hand, the heavenly gods are the lords of the stars.... The classic paradigm of this fairly

> abstract form of deity-formation is the highest conception of the ancient Hindu pantheon, Brahma, as the lord of prayer. Just as the Brahmin priests monopolize the power of effective prayer.
>
> —Max Weber, *The Sociology of Religion*, 1971: 13-14.

Period of Changes

While Hinduism is the religion of the land, there is no one pan-Indian form of religion. In fact, Hinduism has been evolving through the ages. Moreover, it has interesting regional forms and each cultural region has its own distinguishing traits expressed in rituals, customs, ceremonies, festivals and social practices. An early form of Hinduism, based on the worship of mother goddess gradually changed into a religious form commonly referred to as Vedic Hinduism (Box). The advent of Vedic religion led to drastic modification of the ancient faith around the middle of the second millennium B.C. While the Vedic religion remained as a superstructure, regional forms of faith based on a blending of local varieties with the newly introduced elements continued as a substratum.

It is commonly known that the transformation of early Hinduism into Vedic form of religion started with the advent of the Indo-Aryan language, particularly Sanskrit, which became the vehicle for this change. The Vedic theology according to one interpretation "begins with the worship of the things of heaven, and ends with the worship of the things of earth" . The gods of heaven included Surya, Savitar and Bhaga, which really implies the worship of sun in different forms.

Then was the worship of the gods of the intermediate sky, Indra, and earth-born gods, Agni, Soma, etc. Gradually, the concept of *sraddha*, or periodical feast of the dead, led to the emergence of the cult of sacrifice. The Rigveda recorded the practices and the rituals associated with the different forms of worship as the Vedic religion spread to other regions of India from its homeland in the Sapta-Sindhava region. It acquired distinct regional forms. The Hindu pantheon recognized a Super Being, an *Atman* or *Parmatman* which permeated all beings, including the gods they worshipped. The Vedic theology was not a fixed set of ideas. In fact, it continued to evolve from the period of the *Vedas* to that of the *Brahmanas* and the *Upanisads*. During these evolutionary stages significant changes were registered in the Hindu philosophy and the very concept of gods underwent a change.

As Brahmanism became rigid protest movements against the ascetic fraternity of the Brahmans were launched. These protest movements eventually led to the emergence of new faiths, such as Buddhism and Jainism. Buddhism was a heretical movement. Basically, it was a reformation movement and a protest against the ascendancy of the priestly class. Buddhism, in its original form, was not a religion as such. It was simply a monastic organization. It insisted on a non-Brahmanic order. In its historical context, it appears that it was a culmination of the on going conflict between the Brahman and the Kshatriya. In the background of Magadha one can understand that the Kshatriya was in a dominating position.

The fact that these protest movements originated in Magadha explains the nature of the caste conflict so peculiar of the region. Early Buddhism tried to disentangle people from the Brahmanical cult. Among the basic elements of Buddhism were the sanctity of animal life *(ahimsa)* and the craving for salvation *(nirvana)*. However, the early Buddhist gospels were a continuation of the old Hindu beliefs and ethics. Buddha laid an extraordinary stress on the issue of *nirvana*. It is a remarkable feature of India's religious tolerance that in spite of its opposition to Brahmanical supremacy Buddhism co-existed with Brahmanism for more than a millennium.

In its early days Buddhism benefited from its practice of

holding *sangha* (congregation of monks or monastic order) in the propagation and consolidation of Buddhist faith. Through these congregations Buddhist preachers redefined their position vis-a-vis the basic philosophical issues. It is also believed that Buddhism also benefited when the state adopted it perhaps as a state religion under Asoka.

Asoka's role in the propagation of Buddhism to other parts of the world, such as Sri Lanka and Central and Eastern Asia, is too well known. With the passage of time Buddhism got divided into two schools of thought—Hinayana and Mahayana. The Mahayana school which developed under Kanishka promised salvation to the entire universe as against the Hinayana concept of salvation of the few. It is a known history that Buddhism started declining in its homeland while it flourished abroad. Hiuen Tsang's visit to India during 629-45 A.D. confirmed its general state of decay.

It is understood that the main cause of this decline was Brahmanical revivalism attended by a reformation movement within Brahamanism. The rise of Brahamanism is attributed to the efforts of Kumarila Bhatta, a Brahman preacher of Bihar and his disciple, Sankaracharya. There were other factors too.

For example, the Buddhist creed was based on a high standard of morality, much higher than the level of common people. Moreover, rejection of the sacrificial cult and abstraction of the concept of god acted as distracting factors, hastening the decay of Buddhism. The common Hindu was more interested in gods which were concrete objects of worship. Thus, the decline of Buddhism may be an outcome of the resurgence of Brahamanism as well as the natural decay of the Buddhist faith among the masses.

The ascetic fraternity of the Brahmans led to another protest movement which came to be known as Jainism. The movement emerged in the lifetime of Buddha himself. Like Buddha, its leader Vardhamana, popularly known as Mahavira Jina, also came from Magadha. The Brahman authority was so strong that these protest movements gained popularity. Jainism also aimed at *nirvana,* although the Jain concept of *nirvana* was a little different than in Buddhism. Within Jainism a division between Svetambara and Digambara took place at an early stage.

Svetambara originated in the northern and the western region of the country while Digambara had its base in the south. The Digambaras followed the principle of nudity. A third sect which consisted of Dhondias also emerged. Jainism received state patronage under the Chalukyas of Gujarat and Marwar. With this began the process of construction of famous Jain temples in different regions of the country. Unlike Buddhism, Jainism did not suffer a setback. The Jain order survived without much of a conflict, external or internal.

Today, the main Jain sanctuaries are situated in the isolated hills, such as Parasnath in Bengal, Palitana in Kathiawar and Mount Abu in Rajasthan. The census records show a successive decline in the numerical strength of Jain population. It may also be noted that the home of Jainism today lies in Gujarat and adjoining Marwar region of Rajasthan where the followers have been drawn mostly from among the trading communities, and not in Bengal, Magadh or Orissa where the Jain faith originated.

Reform movements, initiated by Buddha and Mahavira, resulted in new formulations of ritual as well as social customs. There was a phase, although short-lived, in Indian history when the tide of Buddhism seemed to submerge the Hindu faith. Buddhism had its sway over a vast region from Bengal to Kashmir and from the Himalayas to Kaveri till a new Brahmanical upsurge brought an equally spectacular change.

The protest movements represented by Buddhism and Jainism also led to reformation within Brahmanism. These reform movements eventually resulted in the emergence of a new form of Hinduism which is currently in practice in all parts of India. Its basic elements consist of use of one sacred language (Sanskrit), through which the ritual is practised, pilgrimage to holy places and the rule of a priestly class which determines the life of the ordinary Hindu men and women. The religion of the peasant, although simpler in its theological content, is equally controlled by the Brahamanical authority.

In the course of time, two sects, *viz.*, Vaisnavism and Saivism emerged among the Hindus. While the former focuses on the worship of Vishnu, the latter on the worship of Siva. Each sect has its own order of rituals performed on specific occasions. However, both the gods are worshipped by the common Hindu folk throughout the country. Both Siva and Vishnu are mentioned in the *Mahabharata* as two separate gods, and at the same time each represented by the other. The Hindu trinity of gods consists of both of them as well as a third god, Brahma, who was treated as the chief of the gods in the epic period. The modern belief is that while Brahma occupies a central position, he is represented by both Vishnu and Siva and in their worship Brahma is also worshipped indirectly.

As pointed out earlier, a remarkable feature of Hindu society is a spirit of tolerance and accommodation of multiple forms of faith. It was because of this spirit of tolerance that other faiths which came from outside flourished. Christianity and Islam came from outside and the people embraced these faiths without much of a social conflict. This has been a historic tradition that successive penetration by non-Indic religions was not resisted. Political conflicts were there between the local kingdoms and the invaders, but they did not result in religious conflicts.

Gradually, these religious faiths created for themselves niches in the Indian social space. As indicated earlier, among the Semitic religions, the earliest to make an impact in India was Christianity. Syrian Christians appeared on the west coast of India in the very first century of the Christian era. Later, in the seventh century A.D., Arab traders brought the message of Islam to the people living on the west coast of India. This happened much before the Muslim (political) conquest of northern India. It is thus obvious that the religious composition of the Indian population has been changing with conversions from one faith to the other. The spatial patterns of distribution of different religious groups were also greatly modified by large scale migration following the partition of India in 1947. Partition brought about a significant change in the distribution and relative strength of different religious faiths in north western, northern and north eastern India.

Geographical Dispersion

The different religious communities of India include major groups, such as the Hindus, Muslims, Christians, Sikhs, Buddhists and the Jains, as well as minor religious faiths, such as Jews and Zoroastrians. Moreover, several tribal communities continue to retain faith in tribal religions based on totemism and animism. The Hindus account for 82 per cent of the total population and are the largest religious group in the country. However, in some districts, Muslims, Christians, Sikhs or the Buddhists are more numerous than the Hindus.

Muslims are the largest minority group and account for 12.12 per cent of the total population. Christians constitute 2.3 per cent of the population and the Sikhs 1.94 per cent. Buddhists and Jains are numerically insignificant accounting for 0.7 and 0.5 per cent of population respectively. It may be noted that while Hindus are generally found everywhere, other religious groups are concentrated in a few pockets only (Table).

Religious Composition of Population, 1991

Religious Groups	*Population*	*% of Total Population*
Hindus	687,646,721	82.00
Muslims	101,596,057	12.12
Christians	19,640,284	2.34
Sikhs	16,259,744	1.94
Buddhists	6,387,500	0.76
Jains	3,352,706	0.40
Others*	3,269,355	0.39
No Religion**	415,569	0.05
All India Population	838,567,936	100.00

* Other religious persuasions.

** Religion not stated.

Source: Census of India, 1991, Table C-9, *Religion.*

It may be useful to identify the main patterns in the spatial distribution of the major religious groups in order to understand the role of religion in defining the parameters of Indian social space and the socio-cultural regionalism based on it.

The Hindus : Hindus are numerically predominant everywhere. However, they are not merely a demographic phenomenon. They have exercised an abiding influence on the sociology and the cultural ecology of all other religious groups which have been derived from Hinduism at different stages of their evolutionary history. The intertwined way of give-and-take between Hinduism and other faiths, such as Christianity and Islam, has given Indian culture a uniqueness of its own. The common features of culture and tradition are shared by all religious groups.

Numerical Strength : With a total population of 687.6 million, Hindus constitute the largest religious community. The Hindu population registered a net increase of 122 million during the 1981-91 decade. There are no specific spatial patterns which can be highlighted in the distribution of Hindu population. In fact, the way the Hindu population is distributed is not different from that of the total population. Two-fifths of the Hindu population lives in the Hindi-speaking northern states (*viz.*, Haryana, Delhi, Uttar Pradesh, Bihar, Rajasthan and Madhya Pradesh).

The two states of Uttar Pradesh and Bihar alone account for more than 27 per cent of the country's Hindu population. Likewise, the four southern states accommodate one-fourth of the Hindu population. In contrast, the northwestern states of Jammu and Kashmir, Himachal Pradesh, Punjab and Haryana have less than 4 per cent of the Hindu population. Similarly, Hindus in the northeastern states of Assam, Arunachal Pradesh, Nagaland, Manipur, Mizoram, Tripura and Meghalaya constitute less than 3 per cent of the country's aggregative Hindu population (Table).

Distribution of the Hindus, 1991

States/UTs	*Population*	*Concentration Index (%)*
States		
Andhra Pradesh	59,281,950	8.62
Arunachal Pradesh	320,212	0.05
Assam	15,047,293	2.19
Bihar	71,193,417	10.35
Goa	756,621	0.11
Gujarat	36,964,228	5.38
Haryana	14,686,512	2.14
Himachal Pradesh	4,958,560	0.72
Jammu & Kashmir	NA	NA

Contd...

States/UTs	*Population*	*Concentration Index (%)*
Karnataka	38,432,027	5.59
Kerala	16.668.587	2.42
Madhya Pradesh	61,412,898	8.93
Maharashtra	64,033,213	9.31
Manipur	1,059,470	0.15
Meghalaya	260,306	0.04
Mizoram	34,788	0.01
Nagaland	122,473	0.02
Orissa	29,971,257	4.36
Punjab	6,989,226	1.02
Rajasthan	39,201,099	5.70
Sikkim	277,881	0.04
Tamil Nadu	49,532,052	7.20
Tripura	2,384,934	0.35
Uttar Pradesh	113,712,829	16.54
West Bengal	50,866,624	7.40
Union Territories		
Andaman & Nicobar Islands	189,521	0.03
Dadra & Nagar Haveli	132,213	0.02
Chandigarh	486,895	0.07
Daman & Diu	89,153	0.01
Delhi	7,882,164	1.15
Lakshadweep	2,337	0.00
Pondicherry	695,981	0.10
All India Total	687,646,721	100.00

Source : Census of India, 1991, Table C-9, *Religion*.

Spatial Distribution : A demography of religions may be helpful in understanding the dynamics of population of different religious groups. It may also be useful to evaluate the spatial patterns in the distribution of religious groups. As noted above, in terms of numerical strength the followers of Hindu religion enjoy a dominating position. At the all-India level they constitute as much as 82 per cent of the total population of the country. Seen in this context, the followers of all other religious persuasions put together have a relatively insignificant share in overall population—just 18 per cent.

Despite this insignificant position, the question of religious minorities continues to be a vexed problem in India. The minority-majority relations have often resulted in social and political problems with far-reaching implications for the civil administration. The Hindu proportion is higher in the rural segment of population—the national average being 84 per cent—thus revealing the predominantly rural character of the Hindu population.

As against this, the proportion of Hindus in the urban population is around 76 per cent. Barring the peripheral areas of the country and a few pockets in the interior, Hindus remain a dominant religious group everywhere in the country. However, in the peripheral areas Hindus are outnumbered by the Muslims (*e.g.*, Jammu and Kashmir), Sikhs (e.g., Punjab), Christians (e.g., Mizoram, Nagaland) and Buddhists (e.g., Arunachal Pradesh). In some regions of the country, such as parts of Orissa, Madhya Pradesh and Andhra Pradesh, Hindus account for more than 97 per cent of the total population.

The proportion of Hindu population is equally high—above 95 per cent—in the sub-Himalayan districts of Uttar Pradesh and Himachal Pradesh. The proportion of Hindus in population remains well above 90 per cent in eastern Madhya Pradesh, eastern Gujarat, southern Karnataka, Tamil Nadu and coastal Andhra Pradesh.

On the other hand, in some districts along the west coast Hindu percentages are significantly low—below 70 per cent. Similarly, in the tribal areas of the northeast, such as Meghalaya, Nagaland and Mizoram, their proportion is less than 20 per cent, and in some cases even less than 5 per cent. Hindus are also a small minority in the Kashmir valley, so much so that in some districts their proportion is less than 2 per cent. In several districts of the northern states of Uttar Pradesh, Haryana and Bihar which happen to be the areas of Muslim concentration, the Hindu proportion goes down to less than 60 per cent. In some districts their share in population is less than 50 per cent.

The regions in which the Hindus are overwhelmingly dominant may be grouped into the following categories:

(a) Regions of tribal concentration and those in the interior of the peninsula where the tribes have been largely Hinduized. These regions are situated away from the entry points through which Christians or Islamic influences percolated (e.g., Orissa highlands, adjacent parts of Andhra Pradesh and the Vindhyan complex in mid-India) thus leaving the original Hindu population unaffected by these developments.

(b) Regions in the sub-Himalayan parts of the north and the northwest (e.g., Himachal Pradesh and Uttar Pradesh hills) where Muslim influences were feeble.

(c) Regions in the Aravallis and the Indo-Gangetic Divide comprising parts of the undivided Punjab. These regions had a high proportion of Hindus in the residual population after the exodus of Muslims in the wake of Partition.

(d) Regions in the Chambal badlands as well as in the former princely states of central India.

(e) Regions, such as the former Rajput princely states of Sirohi, Dungarpur and Banswara (Table).

India: Districts with the Hindu Proportion remaining over 94 per cent

	Type of Region	*Hindu Proportion*	
		Above 98 per cent	*94-98 per cent*
(a)	Regions of tribal	Srikakulam, dominance or those lying away from the northwestern entry points	Visakhapatnam, East Vizianagararn, Godavari, Panch Mahals, Bastar, Bolangir, Jhabua, Batul, Narsimhapur, Dhenkanal, Mandla, Bilaspur, Kalahandi Rajnandgaon, Durg, Raipur, Sambalpur, Keonjhar, Baleswar, Cuttack, Koraput, Ganjam, Puri, Salem, Periyar, Dharmapuri, Dadra and Nagar Haveli, Yanam
(b)	Isolated sub-Himalayan hills	Hamirpur, Mandi, Uttar Kashi, Chamoli, Tehri-Garhwal, Pithoragarh, Almora	Kangra, Bilaspur, Kullu, Shimla, Solan, Garhwal
(c)	Indo-Gangetic divide region of the old Punjab with residual population after Muslim exodus to Pakistan	Rohtak, Rewari, Mahindergarh, Bhiwani	Sonepat, Jind
(d)	Peninsular interior		Mandya, Morena, Tikamgarh, Chhattarpur, Panna, Satna, Rewa, Shahdol, Sidhi, Sonbhadra
(e)	Former Rajput princely states		Sirohi, Dungarpur, Banswara

Source : Based on Census of India, 1991.

Similarly, areas in which the Hindus have been reduced to a minority may be categorized as follows:

(a) Regions of tribal dominance in the Northeast where the tribes have been Christianized.

(b) Regions on the northwestern periphery, such as Jammu and Kashmir and Punjab where either the Muslims or the Sikhs are concentrated.

(c) Regions of sizeable Muslim population, such as Rohilkhand and the upper Ganga-Yamuna doab, or parts of West Bengal and Assam which have been traditional nodes of Muslim convergence (Table).

India: Districts with Hindu Proportion below 21 per cent

15-21 per cent	*10-15 per cent*	*5-10 per cent*	*Below 5 per cent*
Tawang, East Kameng, Senapati, East Khasi hills, West Garo Hills, Amritsar, Sangrur, Faridkot	Upper Subansiri, Chandal	West Siang, Churachandpur, Ukhrul, East Garo Hills, Phek, Mokokchung, Mon	Tamenglong, Jaintia Hills, West Khasi hills, Aizawl, Lungloi, Chhimtuipui, Zunheboto, Wokha, Tuensang, Lakshadweep

Source: Based on Census of India, 1991.

The Buddhists : At the 1991 census a total of 6.38 million Buddhists were enumerated. They constituted 0.76 per cent of the country's population. The pockets of traditional Buddhism are located in Ladakh, Himachal Pradesh, Sikkim, Mizoram and Tripura. Buddhism which was once holding sway over many regions of the country gradually declined in following.

Some Buddhist following continued in the peripheral regions of the country. The bulk of the Buddhist population today lives in Maharashtra (as much as 79 per cent). These neo-Buddhists are converts from Hindu low castes. They were initiated into Buddhism by Baba Saheb Ambedkar in protest against the high caste domination. In many districts of Maharashtra these neo-Buddhists account for 5-17 per cent of the population. The Buddhist proportion is significantly high—more than 12 per cent—in Akola, Nagpur, Wardha, Buldana, Amravati, Chandrapur, Parbhani, Bhandara and Gadchiroli. From 5 to 10 per cent of the population consists of Buddhists in Nanded, Yavatmal, Jalna, Aurangabad, Ratnagiri, Greater Bombay and Satara. The Buddhist proportion is very high (above 60 per cent) in the North Sikkim and significantly high in the East, South and West Sikkim. More than one-tenth of the population in South Tripura and Darjiling consists of the Buddhists. There are two main pockets of the Buddhist concentration in Mizoram. They are situated in Chhimtuipui and Lunglei districts. In the former more than 28 per cent of the district population consists of Buddhists. In the latter their share is about 16 per cent. Another pocket of the Buddhist concentration is found in Arunachal Pradesh where the Buddhists constitute about 80 per cent of population in Tawang. About one-half of the population in West Kameng district is Buddhist in its religious persuasion. The Buddhists, presence is also quite significant in Changlang district. About three-fifths of the population of Lahaul and Spiti in Himachal Pradesh consists of Buddhists. Kinnaur also has a significant proportion of Buddhists in its population (Table).

Frequency Distribution of Districts with Sizeable Buddhist Population

Percentage Categories	*No. of Districts*	*Names of Districts*
Above 60	2	Tawang, North Sikkim
40-60	2	Lahaul and Spiti, West Kameng
2040	6	Changlang, Kinnaur, Chhim-tuipui, East Sikkim, South Sikkim, West Sikkim
10-20	14	Lohit, Parbhani, Nanded, Buldana, Akola, Amravati, Wardha, Nagpur, Bhandara, Chandrapur, Gadchiroli, Lunglei, South Tripura, Darjiling
Below 10	8	Dibang Valley, North Tripura, Greater Bombay, Ratnagiri, Satara, Aurangabad, Jalna, Yavatmal

Note: Districts in which the Buddhist proportion in population is less than 5 per cent have been ignored. *Source:* Based on Census of India, 1991.

The Jains : About 60 per cent of the country's 3.27 million Jains live in the three states of Rajasthan, Gujarat and Maharashtra. The Jain population is largely concentrated in the urban centres of these three states as well as in the neighbouring states of Madhya Pradesh and Uttar Pradesh.

Their preference for urban areas is, however, understandable in view of the fact that the Jains are mainly a trading community. Even in small towns they control petty business. The proportion of the Jains in the population of these states is, however, insignificant. For example, the Jains account for just 1.28 per cent of the total population in Rajasthan, 1.22 per cent in Maharashtra and 1.19 per cent in Gujarat. As against this their proportion in the urban population of these states is high—3.4 per cent in Rajasthan, 3.5 per cent in Maharashtra and 2.6 per cent in Gujarat.

A perusal of Table reveals that the Jains have favoured their concentration in the dry regions of Rajasthan and Gujarat. The two districts in which their share is the highest in urban population are Barmer (13.4 per cent) and Jalor (10.5 per cent). These two districts apart Jains generally constitute from 2-10 per cent of the urban segment of the population in a cluster of districts in Rajasthan, Gujarat, Madhya Pradesh and Maharashtra. In another cluster of adjacent districts their proportion varies between 5 and 10 per cent.

These districts lie in the western and northern parts of Madhya Pradesh, Bundelkhand region of Uttar Pradesh and southern and eastern parts of Rajasthan. In no less than 30 districts distributed over Rajasthan, Gujarat, Madhya Pradesh, Maharashtra and western Uttar Pradesh Jains account for 2-5 per cent of urban population. Elsewhere in India they are conspicuous by their absence (Table).

Frequency Distribution of Districts according to the Proportion of the Jains in Urban Population, 1991

Percentage Categories	*No. of Districts*	*Names of Districts*
Above 10	2	Barmer, Jalor (Rajasthan)
5-10	15	Lalitpur (Uttar Pradesh), Udaipur, Banswara, Dungarpur, Chittorgarh, Tonk, Sirohi (Rajasthan), Surendarnagar (Gujarat), Guna, Sagar, Damoh, Mandsaur, Ratlam, Jhabua, Vidisha (Madhya Pradesh)
2-5	30	Bhavnagar, Kachch, Banas Kantha, Sabar Kantha Gujarat), Bikaner, Bhilwara, Bundi, Jhalawar (Rajasthan), Firozabad, Mainpuri (Uttar Pradesh), Shivpuri, Tikamgarh, Panna, Ujjain, Shajapur, Dhar, Indore, Sehore, Raisen, Jabalpur, Narsimhapur, Seoni, Balaghat, Rajnandgaon (Madhya Pradesh), Greater Bombay, Ahmadnagar, Pune, Satara, Sangli, Kolhapur (Maharashtra)

Source : Based on Census of India, 1991.

The Christians : Christians are among the major religious minorities in India today. They may perhaps be treated as a unique religious group on several grounds. First, like the Muslims, they are widely distributed in the country forming a sizeable proportion of population in many states and union territories. Secondly, Christianity was introduced by European missionaries from the very early stages. It, therefore, became a powerful vehicle of social and cultural change in the country.

It brought in its wake a series of changes popularly expressed by the term 'Europeanization'. Thirdly, since the missionaries generally worked among the poor and the most deprived sections of the Indian population, Christianity indirectly contributed towards the alleviation of poverty and the upliftment of the down-trodden masses. Fourthly, Christian missionaries had a wide sweep and were in a position to reach the most isolated and the backward parts of the country. As a result Christian converts were drawn from diverse ethno-lingual groups. Christianity thus acted as a bridge between diverse elements within India's tribal and caste-based social order. Lastly, the European connection was important. It was through this channel that financial resources could flow into India for social work of great local relevance. In fact, the pioneering work that the Christian missions initiated in the field of education, health and social rehabilitation remains a landmark. It has no parallels in history.

Advent of Christianity in India : Christianity spread in India in several phases. The beginnings are traceable in the establishment of the Syrian Church in Malabar, supposed to have been founded by St. Thomas, one of the 12 apostles of Jesus. The common belief is that he preached in India. The belief, however, remains unsubstantiated. Another development which took place in 345 A.D. was the arrival of a Christian missionary, Bishop of Edessa, also named as Thomas. He came from Iran and started preaching in South India. The Syrian Church was already active in India at the beginning of the sixth century A.D. It formed a branch of the Nestorian Order. Christian presence in India was also noted by Marco Polo, who visited India in the thirteenth century A.D. Christianity received an added impetus with the arrival of the Portuguese in the fifteenth century. Soon the Franciscan missionaries started their work in India with Goa being recognized as Bishopric in 1534. Eight years later St. Francis Xavier reached Goa.

As in the case of the earlier missionaries his work remained confined to the Malabar coast and the southern districts of what is now known as Tamil Nadu. So far converts to Christianity were Europeanized so much that there was a total break from the Indian cultural tradition. This, however, proved to be a setback. This *modus operandi,* changed with Robert de Nobili (died 1605) who settled in Madurai and allowed the Christian converts to retain their cultural practices even after conversion. His method of work was so appealing that even the high caste Hindus converted to Christianity.

The French who came by and large the same time as the Portuguese helped propagate Catholic faith. While the Portuguese were having their sway on the west coast, the French founded their early settlements on the East coast, operating mainly in the region lying to the south of Madras (now Chennai) particularly from their enclave in Pondicherry. The net result was that the efforts of the Portuguese and the French helped in the propagation of Christianity.

The Catholic church could strike its roots in the southern region from coast to coast as the land was dotted with churches, schools and hospitals. The missionaries picked up the local languages and this added to their effectiveness in the missionary work. The Catholic population multiplied with multi-pronged social welfare activity serving mostly the deprived and the down-trodden sections of Indian population.

The first Protestant Mission was established in 1705 by the Danish Lutherans. A Baptist Mission was established at Serampore in 1792. While the East India Company was not directly involved in missionary work, there were Anglican Chaplains who were employed as Company's functionaries. They took interest in the promotion of Christianity. By 1813 the Company finally granted permission to Anglican bodies for missionary work. As a result, the first Anglican Bishopric was established at Madras. The East India Company also agreed to the foundation of the Episcopal See at Calcutta. Both Catholic and Protestant Missions continued their work during the nineteenth century. However, the Catholics were largely confined to the southern region of India. The missionary work was also carried out by the American Board of Foreign Missions in the early decades of the nineteenth century. Their efforts were mainly focused on the United Provinces of Agra and Oudh as well as Punjab. Field stations were established at important towns in these provinces during the period 1834-55. The Telugu Mission and Assam Mission, launched in 1840-41, proved to be highly successful (Box).

> ***The Ecclesiastical Organization :*** The Catholic Church has 15 metropolitan areas, each presided over by an Archbishop with several suffragan Bishops in the area. These Metropolitan Sees include three of the Syrian Rite, those of Ernakulam, Changanacherry and Trivandrum. The number of Bishoprics is over 60. The former Anglican Church is now known as the Church of India, Pakistan, Burma and Ceylon. The Metropolitan resides in Calcutta and there are 13 Episcopal Sees under him. The Methodist Episcopalians have four residing Bishops in India. The Metropolitan, or Catholics, of the Jacobite Syrian Church resides in Kottayam and has nine Bishops under him. There are four other independent Bishops of the Syrian Rite in Kerala.
>
> — *The Gazetteer of India*, Indian Union, Vol. I, 1965, Country and People: 497

The fact that the Roman Catholics constitute a major chunk of the Indian Christian community is also evident from the number of Catholic institutions, such as churches, schools, colleges and hospitals. The church plays an important role in the social integration of the community. The distribution of Catholic church in India, related as it is with the concentration of Catholic population, appears to be an outcome of the long history of diffusion of Christianity.

The emergence of Catholics in the north was rather delayed. The progress of Christianity in the north may be related with the British ascendancy. However, the British were mostly Protestants. Most of the Catholic churches in North India are, therefore, mostly recent, even as recent as independence. The spread of Catholic faith in the northern region, initially subdued, left the Protestants behind. The major concentration, however, remains in the south, west and the north-east. The distribution of the Catholic churches follows the same pattern.

> ***Christianity at the Beginning of the Twentieth Century*** : The Christian community now numbers only three millions, of whom more than two-and-a-half million are native converts, and the remainder Europeans or Urasians. Of the native Christians, about two-fifths are Roman Catholics, and one-eighth Romo-Syrian; one-ninth belong to the Anglican body, one-eleventh are Jacobite Syrians, one-twelfth Baptist... Nearly two-thirds of the total Christian population are found in Madras, including the native states of Cochin and Travancore.
>
> In these states, where the Syrian Church is strongest, nearly a quarter of the entire native population profess Christianity. In British territory, it is in the eight southern districts, the scene of the labours of St. Francis Xavier and Swartz, that Christians are most numerous. Then come the district of the Telugu country- Kistna where they are mainly Baptist and Lutherans; Nellore, nearly all Baptists; and Kurnool, Baptist with a respectable minority of Anglicans.
>
> —*The Imperial Gazetteer of India*, The Indian Empire, 1908, Vol. I: 443

As noted earlier, the spread of Christianity owes much to the dedicated social work of missionaries in the area of education and health. The Christian colleges in Indian cities were by far the best institutions of higher learning. The Christian missions established no less than a hundred first-grade colleges, hundreds of high schools and middle schools and thousands of primary and elementary schools in different parts of the country.

Their main focus was on women's education which was grossly neglected in India. As part of their social work the missionaries also established orphanages, homes for the aged and the handicapped, hospitals, dispensaries and leprosy and tuberculosis sanatoria (Box).

Numerical Strength : The 1991 census enumerated 19.64 million Christians in India. The Christian population increased from 10.53 million to 19.64 million in 1991, registering an average annual growth rate of 2.88 per cent.

Distribution of Christians, 1991

States/ UTs	*Population*	*Concentration Index (%)*
States		
Andhra Pradesh	1,216,348	6.19
Arunachal Pradesh	89,013	0.45
Assam	744,367	3.79
Bihar	843,717	430
Goa	349,225	1.78
Gujarat	181,753	0.93
Haryana	15,699	0.08
Himachal Pradesh	4,435	0.02
Jammu & Kashmir	NA	NA
Karnataka	859,478	438
Kerala	5,621,510	28.62
Madhya Pradesh	426,598	2.17
Maharashtra	885,030	431
Manipur	626,669	3.19

Contd...

States/ UTs	*Population*	*Concentration Index (%)*
Meghalaya	1,146,092	5.84
Mizoram	591,342	3.01
Nagaland	1,057,940	5.39
Orissa	666,220	3.39
Punjab	225,163	1.15
Rajasthan	47,989	0.24
Sikkim	13,413	0.07
Tamil Nadu	3,179,410	16.19
Tripura	46,472	0.24
Uttar Pradesh	199,575	1.02
West Bengal	383,477	1.95
Union Territories		
Andaman & Nicobar Islands	67,211	0.34
Dadra & Nagar Haveli	2,092	0.01
Chandigarh	5,030	0.03
Daman & Diu	2,904	0.01
Delhi	83,152	0.42
Lakshadweep	598	0.00
Pondicherry	58362	0.30
All India Total	19,640,284	100.00

Source : Census of India, 1991.

The Christian population reveals a strong tendency of concentration in the southern states of Kerala, Tamil Nadu, Andhra Pradesh and Karnataka. This is, however, in conformity to or with the historical processes through which Christianity spread in the country. About 56 per cent of the total Christian population of the country lives in the four southern states alone.

Kerala alone has almost one-third of the country's Christian population, while Tamil Nadu takes care of about one-sixth of the national total. A second major concentration of the Christians is seen in the northeastern states which together share 22 per cent of the total Christian population of the country (Table).

As noted earlier, the Christian population consists of Catholics as well as Protestants. Subsumed in the Christian population are more than half a million orthodox Syrian Christians and more than a lakh Anglo-Indians consisting of Protestants as well as Catholics. At the 1901 census nearly 3 million Christians were enumerated. Of these 2.5 million were native converts and the remainder Europeans or Eurasians.

Spatial Distribution : A perusal of the district-level data shows pockets of Christian concentration in the southern state of Kerala where as many as seven districts have a Christian proportion ranging between 20 and 50 per cent. The Christian proportion is particularly high in Kottayam (45.83 per cent), Idukki (42.23 per cent) and Ernakulam (73.83 per cent). In all these districts Christians have a sizeable population in both the rural and urban segments of population.

Gumla in the Chhotanagpur region of Bihar, North Goa and South Goa districts and Kanyakumari also fall in the same category. On the periphery of these pockets of high concentration in Kerala and Tamil Nadu the Christian population thins out although their share remains between 10 and 20 per cent.

The proportion of Christians in population remains above 50 per cent in as many as 20 districts situated mostly in the northeastern states of Manipur, Meghalaya, Mizoram and Nagaland. However, while their share in population is high—in some districts above 90 per cent—their numerical strength is evidently low.

In Mizoram, for example, their share in the total population is 83.8 per cent, followed by Nagaland with 80.2 per cent of its population consisting of Christians. The Christian proportion also remains high in Meghalaya and Manipur. Goa has a sizeable Christian population with at least one-third of its population being Christian. Outside these pockets of high concentration one notices small Christian populations in a number of districts in Uttar Pradesh and Punjab. Gurdaspur in Punjab is a notable example (Table).

Frequency Distribution of Districts According to the Percentage of Christians in Total Population, 1991

Percentage Category	*Number of Districts*
Above 90	11
80-90	3
50-80	6
20-50	13
10-20	10
5-10	27
2-5	46
1-2	31
Below 1	296

Source : Based on Census of India.

The Muslims : Like the Christians, Muslims are an important web in the social fabric of India. While India is the original home of Hinduism which constitutes the system of beliefs and rituals for the great majority of our people, Indian ethos has liberally accommodated Islamic elements of religion, philosophy and culture which penetrated the sub-continent at different points of time in history. There is evidence to show that this give-and-take continued for centuries, and contributed to the distinguishing traits of India's composite culture. The Muslims of India are integrated with other sections of the Indian population—Hindus, Buddhists, Sikhs and Christians—at different levels of regional hierarchy. Responding to distinctly different environmental conditions, assimilating diverse ideas in the historical process of cultural intermingling and reaching disparate levels of socio-economic development, Muslims of India, like other communities, have developed their cultural traits under distinct regional moulds. It is in the context of these regional facets that the basic parameters of their identity can be defined.

Advent of Islam in India : As noted earlier, Indian ethos has accommodated elements of other faiths which reached the Indian sub-continent at different points of time in history. The earliest among these faiths to make an impact on the Indian people was Christianity. Several centuries later, Islam was introduced to the people living on the west coast of India by the Arab traders and this happened much before the Arab conquest of Sind. The origins of Muslim society in India may be traced in the chequered history of India's contact with Islam. The following major phases may be identified:

First, Arab trading settlements which already existed on the west coast of India emerged as centres of dissemination of information about the Muslim faith, a few years after the Prophet's proclamation of his mission in the seventh century A.D. Secondly, Arab sea-farers who had trade contacts with the people in the coastal regions of Gujarat and Baluchistan, either single-handedly or in cooperation with Persians, also became carriers of the message of Islam by around the same time. Thirdly, Arab conquest of Sind in the early years of the eighth century set the stage for the political ascendancy of Islam in the lands lying to the west of Indus.

The process culminated in the establishment of Delhi Sultanate in 1192. Fourthly, northern and central parts of India from Kabul to Bengal and from Kashmir to Tapti (now Tapi) were annexed by successive empires under the Tughlaqs, Khiljis,

Lodhis and the Mughals. Fifthly, while the powerful Mughal empire crumbled after the death of the last emperor in 1707, residuals of the Mughal dynastic rule survived in many parts of India, particularly Awadh, Bihar, Bengal and central India.

Sixthly, many of these centres of local power were recognized as princely states by the British who helped them survive as islands of Muslim feudal order till independence in 1947. Lastly, while the political fortunes of the ruling elites—Turks, Afghans, Mughals—fluctuated, Sufis established their *silsilahs* in different parts of India as a parallel development. This continued as a phenomenon throughout the medieval period. The Sufi *khanqahs* often became centres of dissent against the despotic Muslim rulers. The Sufi message of love, compassion, peace and universal brotherhood appealed the masses cutting across the barriers of religion, caste and creed. This paved the way for a better understanding of the Islamic faith, eventually leading to conversion to Islam.

Keeping in view the nature of influences exercised from different directions at different points of time as outlined above, it is possible to identify the following ethnic roots of the Muslim community of India:

(a) Malabar and adjoining parts of the Deccan, such as the dominions of the erstwhile Nawab of Hyderabad, were exposed to the Arab ethnic influences.

(b) Hyderabad became a centre of intermingling of Telugu culture with the North Indian culture. The latter had already found a base in the Deccan environment.

(c) Gujarat, Makran sea-coast and parts of coastal Maharashtra came under the influence of Arabs as well as Persians.

(d) Sind, Punjab and other parts of northern India received layers after layers of Arab, Afghan, Turkman (Turkic), Uzbek and Persian influences leaving behind their traces in the ethnic composition of the Muslim community living in northern, northwestern and central parts of India.

(e) These exogenetic influences imply only the nature of initial contacts. The factual position is that the Muslim masses were largely derived from the local ethnic groups, both Hindu and Buddhist, and carried with them across the religious boundaries their distinctive ethnic and cultural characteristics even after conversion. This largely explains the ethnic and cultural diversity of the contemporary Muslim society not only in India but also in the neighbouring countries of Pakistan and Bangladesh. Their social structure appears to be the end-product of this long process of miscegenation.

It is generally known that trade relations between the Arabs and the Indians existed much before historical records were maintained. Much before the advent of Islam in Arabia, Arabs had settlements on the west coast of India. Obviously Islam was introduced to the coastal regions of South India within a few years of its rise in Arabia. Most probably the trading settlements started disseminating information about the Islamic faith. Inter-marriages between Arabs and the local people resulted in the emergence of the well known Muslim community of Malabar, called the *Moplahs.* These stray references apart, there is no record of any systematic Muslim missionary effort in South India in the early centuries of Islam.

Evidently, South India was introduced to Islam much earlier than North India. So far as North India is concerned, the Arab conquest of Sind by the young Muslim General, Muhammad Bin Qasim, in 711 A.D., was an isolated affair and affected only the situation in the trans-Indus region. Northern India, however, felt shock waves of Turk and Afghan invasions finally resulting in the establishment of the Sultanate in Delhi in 1192. From 1192 to 1707 different dynasties ruled over India, particularly over the northern region of the country from Kabul to Bengal and from Kashmir to Khandesh.

It may be recalled that the role of the Sufis in the spread of Islam was far more important than the role of the state. Arab chronicles speak of Multan and Uch as centres of missionary activity in the early years of Arab conquest of Sind. Lahore became another centre of Sufi activity by the beginning of the tenth century A.D. A Sufi saint Sheikh Ismail, who came to Lahore in the wake of invasions of Mahmud of Ghazni, established himself as a successful missionary much before a Muslim governor was appointed by Mahmud of Ghazni.

It appears that Gujarat also came under the influence of the Arabs and the Persians at around the same time as South India. It was mainly because the region had a tradition of maritime trade with the Persians and the Arabs. The role of Sufis became obvious in the propagation of Islam with Khwaja Moinuddin Chishti coming to Ajmer while the city was the seat of Hindu power. As founder of the Chishtiyah *silsilah* of Sufis, he played an important role as a missionary.

Numerical Strength : With a total population of 101.6 million, Muslims constitute a little over 12 per cent of the country's population. This figure, however, excludes the Muslims of Jammu and Kashmir where due to operational difficulties the 1991 census count was not made. At the earlier census of 1981, the total Muslim population was 75.5 million. At the 1981 census, Assam was excluded from census operations for the same reason. The two figures are, therefore, not comparable. The data can be made comparable only by excluding the two states of Jammu and Kashmir and Assam from this analysis. After this adjustment, the size of the Muslim population comes to 71.4 million in 1981 and 95.2 million in 1991. The Muslim population, thus, registered a net increase of 23.8 million during the 1981-91 decade, recording a decadal growth rate of 32.7 per cent or an average annual exponential growth rate of 2.83. This was by far the highest growth rate registered by any religious community during the decade. Such a high growth rate of the Muslim population was found to be alarming by many and naturally it evoked wide ranging reactions from different sections of the Indian society.

About one-fourth of the 101.6 million Muslims enumerated at the 1991 census were found in Uttar Pradesh, about one-sixth in West Bengal and about one-eighth in Bihar. The three states together shared amongst themselves more than one-half of the total Muslim population. If Assam is added to this list, the four states together make up for three-fifths of the national total.

This high concentration of Muslims is in conformity with the general pattern of population density in the country. It is also in conformity with the historical processes which contributed to the spread of Islam in this country. The Muslim population is also sizeable in the states of Maharashtra, Andhra Pradesh, Karnataka and Kerala. These four states together share one-fourth of the country's Muslim population. Thus, more than 84 per cent of the Muslim population is concentrated in the eight states of the Indian Union. These patterns of distribution indicate that there are two major clusters: (i) North Indian cluster, encompassing Uttar Pradesh, Bihar, West Bengal and Assam; and (ii) South Indian cluster, consisting of Andhra Pradesh, Karnataka and Kerala. Maharashtra appears to be transitional to the southern cluster (Table).

Distribution of Muslims, 1991

States/ UTs	*Population*	*Concentration Index (%)*
States		
Andhra Pradesh	5,923,954	5.83
Arunachal Pradesh	11,922	0.01
Assam	6,373,204	6.27
Bihar	12787,985	12.59

Contd...

States/ UTs	*Population*	*Concentration Index (%)*
Goa	61,455	0.06
Gujarat	3,606,920	3.55
Haryana	763,775	0.75
Himachal Pradesh	89,134	0.09
Jammu & Kashmir	NA	NA
Karnataka	5,234,023	5.15
Kerala	6,788,364	6.68
Madhya Pradesh	3,282,800	3.23
Maharashtra	7,628,755	7.51
Manipur	133,535	0.13
Meghalaya	61,462	0.06
Mizoram	4,538	0.00
Nagaland	20,642	0.02
Orissa	577775	0.57
Punjab	239,401	0.24
Rajasthan	3,525,339	3.47
Sikkim	3,849	0.00
Tamil Nadu	3,052,717	3.00
Tripura	196,495	0.19
Uttar Pradesh	24,109,684	23.73
West Bengal	16,075,836	15.82
Union Territories		
Andaman & Nicobar Islands	21,354	0.02
Dadra & Nagar Haveli	3,341	0.00
Chandigarh	17,477	0.02
Daman & Diu	9,048	0.01
Delhi	889,641	0.88
Lakshadweep	48,765	0.05
Pondicherry	52,867	0.05
All India Total	101,596,057	100.00

Source: Census of India, 1991.

Spatial Distribution : The distribution of Muslim population in India today is the result of a process extending over almost a millennium. While it is not intended here to analyze this process in depth, it is necessary to indicate the factors which have played a role in determining the current distribution patterns. These factors are enumerated below:

(i) The place of origin of Muslims outside the sub-continent and the routes taken by them—both land and sea—have greatly influenced the density of Muslim population as witnessed in India today. It is evident that the farther we go from the entry points and the routes of migration, the proportion of the Muslims in the total population decreases. This broadly explains the thinning out of the Muslim population as one moves from west to east or north to south in the undivided India. The concentration of Muslims in regions, such as Bengal, may, however, be explained by relating it to the process of mass conversion rather than to the routes of migration.

(ii) The existing distribution pattern is also the result of the process of mass conversion from Buddhism and Hinduism to Islam that took place on the eastern and northern peripheries of the sub-continent in Kashmir and Bengal. A part of Bengal in which the Muslims were in majority was later separated from India to constitute an eastern wing of Pakistan in 1947. After the war of liberation in 1971, the territory was renamed as Bangladesh. Another pocket of Muslim dominance lies in Kashmir. The origins of Muslim society in Kashmir may also be related to the process of mass conversion facilitated by the Sufi *silsilahs*.

(iii) The centres of political authority in the principalities, states and kingdoms, where political power was in the hands of the ruling Muslim elites acted as centres of attraction for the Muslim masses in the surrounding regions. As a consequence, a peculiar pattern of concentration of Muslim population emerged. Best examples of this phenomenon may be seen in the erstwhile princely states of Hyderabad, Bhopal, Tonk, Junagadh and Rampur.

(iv) The agglomeration of Muslims in urban centres, it appears, has been greatly influenced by the migration of Muslim artisans from villages and small towns to large industrial centres. This happened particularly in industries which were connected with traditional Muslim crafts. Today, a large concentration of Muslim population is seen in the metropolitan areas of Calcutta, Mumbai (formerly Bombay), Ahmadabad, Kanpur and so on. In each of these centres Muslim artisans and industrial workers are generally attached to certain industries, such as textiles and leather goods.

(v) One can also identify a process of spill-over of the Muslim masses from the rural areas with a pressure on agricultural land to pioneer fringes in their neighbourhood. This led to the redistribution of rural population. This may be illustrated by citing the example of East Bengal from where a spill-over of Muslim population to the neighbouring districts of Assam continued throughout the early part of the nineteenth century.

(vi) The mass exodus of Muslims from India in the wake of Partition of the country in 1947 greatly modified the demographic picture as it had evolved through centuries. Muslim exodus from northern India was counter-balanced by an influx of Hindus and Sikhs from the newly-created territory of Pakistan.

(vii) The demographic picture has been further modified since Partition. This redistribution of population caused either by the developmental processes, or communal disharmony, is changing gradually the religious composition of population in many large cities of India. This has often led to the formation of Muslim enclaves within the major cities, such as Kanpur, Ahmedabad, Moradabad, Firozabad and Aligarh. The distribution of Muslim population in India today reveals strong tendencies of clustering and concentration in certain parts of the country, such as Jammu and Kashmir, Uttar Pradesh, Bihar, West Bengal, Haryana, Rajasthan and Kerala. The following pockets of Muslim concentration may be identified:

(a) Parts of Jammu and Kashmir, such as the upper Jhelum valley, the upper Chenab valley (e.g., Kishtwar) and the adjoining areas. In these pockets the Muslims constitute more than 90 per cent of the population. Outside the valley of Kashmir, there is a general thinning out of Muslim population. However, the Muslims constitute two-thirds of the population of Doda and Punch and a little over one-third of the population of Ladakh and Udhampur. In Jammu and Kathua, the Muslim proportion sharply declines, although it remains around 10 per cent.

(b) Delhi-Mewat and the Ganga valley. In this cluster, the Muslim proportion remains significant all over the region from the National Capital Territory of Delhi to Murshidabad and Maldah districts of West Bengal. The following patterns of the second order may be noted:

(i) Murshidabad and Maldah with the Muslim proportions ranging between 46 and 56 per cent.

(ii) The Muslim proportions are quite significant in Rampur (Uttar Pradesh), West Dinajpur

(West Bengal), Purnia (Bihar), Moradabad and Bijnor (Uttar Pradesh).

(iii) These districts of high agglomeration are skirted by districts in which the Muslim proportion generally remains between one-third and one-fifth. The districts in this category include Saharanpur, Bareilly, Muzaffarnagar, Bahraich, Pilibhit, Meerut and Gonda (Uttar Pradesh), Birbhum, Nadia and 24 Parganas (West Bengal).

(c) Brahmaputra and the Barak valleys of Assam particularly Goalpara, Nowgaon, Kamrup and Darrang districts of Assam where the Muslim proportion ranges between 20 and 40 per cent.

(d) Dry regions of Rajasthan and Gujarat, particularly the bordering districts of Jaisalmer, Barmer and Kachch with the Muslim proportion ranging between 10 and 20 per cent.

(e) The peninsular interior including Malwa, Khandesh, Marathwada and other parts of the old princely states of Hyderabad and Mysore. While the Muslims account for one-fourth of the population of Hyderabad, in other districts of the peninsular interior their proportion ranges between 10 and 25 per cent. Districts, such as Ujjain, Indore, East Nimar and Sehore (Madhya Pradesh), Aurangabad, Parbhani, Nanded, Osmanabad and Akola (Maharashtra), Bellary, Bijapur, Dharwar, Gulbarga, Bidar and Raichur (Karnataka), and Cuddapah, Anantapur, Kurnool, Medak and Nizamabad (Andhra Pradesh) may be cited as examples.

(f) Malabar, particularly the coastal districts of Kerala, such as Kozhikode, Cannanore and Palghat, Ernakulam, Trichur, Quilon and Trivandrum (Kerala).

Evidently, the pockets of Muslim concentration conform to the geographic patterns in the density of general population in India (Box).

> ***Density of Muslim Population :*** Even more interesting is the correspondence between the zones of high Muslim concentration and the high population density regions of the country. Muslims have the highest numerical strength in the high density zones of West Bengal, in the adjoining districts of Bihar and in Kerala—24-Parganas, Murshidabad, Purnia and Kozhikode holding the four leading positions. The fertile, alluvial plains of the Ganga and the Brahmaputra valley and the Malabar coast, being the most densely peopled parts of the country, also have the highest Muslim concentrations.
>
> However, if the Muslims are conspicuous by their absence in some areas of high density, such as the Punjab plains, the explanation lies in the distorting effects of the political processes which 'were generated in the recent past. In consonance 'with the general distributional pattern, the density of Muslim population also declines towards the arid/semi-arid west, the Malwa/Central Indian uplands, the peninsular interior and the Ghats.
>
> —Moonis Raza, Aijazuddin Ahmad and Nafis Ahmad Siddiqi in Zafar Imam (ed.), *Muslims in India*, 1975:119.

The district-level patterns lead to several conclusions. First, the concentration of Muslims in the Kashmir valley is significant in many ways. In the valley districts, Muslim proportion generally remains above 90 per cent. Similarly, in the districts surrounding these pockets of high concentration, Muslims remain significant in terms of their numerical strength. To the east of the Kashmir valley lies Doda district 'where the Muslim proportion is quite high. Secondly, the Muslims are also predominant in a large number of districts in the North Indian Plain from Uttar Pradesh to West Bengal, where their proportion ranges between 20 and 60 per cent. Included in this category are the districts of Murshidabad (61.4 per cent), Rampur (47.95 per cent), Moradabad (42.7 per cent), Bijnor (40.35 per cent), Bareilly (32.79 per cent), Pilibhit (23.12 per cent), Saharanpur (36.12 per cent), Muzaffarnagar (34.52 per cent), Meerut (27.49 per cent) and Ghaziabad (21.16 per cent). Thirdly, the share of Muslims in the district population is fairly significant in several other regions, such as Haryana (Gurgaon), Rajasthan, Gujarat, Madhya Pradesh, Andhra Pradesh and Kerala.

While Muslims are a phenomenon in themselves in several parts of India, there are hundreds of other districts in which they are not so numerous. They have been living for ages as an inseparable part of the rural and urban milieu and bound together by the age-old ties in economy, artisan industry, arts, crafts, folk cultures and social customs (Table).

Frequency Distribution of Districts According to the Percentage of Muslims in Total Population, 1991

Percentage Categories	*Number of Districts*
Above 80	Nil
50-80	10
20-50	43
10-20	117
5-10	108
1-5	100
Below 1	59

* Excludes Jammu and Kashmir. *Source:* Based on Census of India, 1991.

Muslims in the Urban Social Space : The above description is restricted to the proportion of Muslims in the overall population. While Muslims are generally a phenomenon in the rural areas, there are certain districts in the north, northwest and the interior of the peninsula where they constitute a substantive proportion of urban population.

Their degree of urbanization is much higher in these regions. In a nutshell, there were more than 36 million Muslims living in the cities and towns of different sizes at the 1991 census: the urban percentage being as high as 35 as against the national average of 26. At the national level Muslims constitute about 16 per cent of the urban population of the country. The degree of urbanization among the Muslims is obviously high. There are several reasons for this high level of urbanization. These reasons are historical as well as those related to the Partition of the country.

The traditional association of Muslims with certain branches of industry, including artisan industry, also led to their agglomeration in selected cities and towns. In the course of time, the capital cities of princely states, some of whom were residuals of Mughal power (*e.g.*, Shahjahanabad, Lucknow, Agra, Murshidabad, Bhopal, Hyderabad, Aurangabad, Junagadh and Rampur) acquired a typical Muslim character. Many of these cities are still prominent as centres of Muslim culture. Because of their traditional association with certain arts and crafts Muslims were attracted to many other cities, such as Varanasi, which has a predominantly Hindu character.

There are several clusters in which the Muslims are an important component in the urban population. First, the dominance of Muslims in the valley of Kashmir and in Kishtwar and Punch may be noted. It has resulted in their agglomeration in urban areas as well. In all the districts of Kashmir valley as well as in Ladakh, Kathua and Doda districts, Muslims constitute an overwhelming proportion of urban population.

Secondly, in western Uttar Pradesh, there are several cities and towns (*e.g.*, Amroha, Sambhal and Rampur) in Rampur and Moradabad districts where the Muslim proportion is around 70 per cent. The Muslim proportion is also very high in Malegaon in Maharashtra (67.4 per cent) and the Garden Reach area of 24 Parganas district in West Bengal.

Thirdly, Muslim percentages remain significantly high (40-55 per cent) in several towns and cities of Cuddapah, East Nimar, Thane, Sikar, Meerut, Moradabad and Shahjahanpur districts. Reference may be made to places, such as Bhiwandi, Burhanpur, Cuddapah and Sikar where 40-55 per cent of the population is Muslim. Lastly, from one-fifth to two-fifths of the urban population consists of Muslims in as many as 83 towns/cities spread over the states of Uttar Pradesh, Bihar, West Bengal, Madhya Pradesh, Gujarat, Maharashtra, Andhra Pradesh, Karnataka, Tamil Nadu and Kerala.

Notable among the urban centres in this category are Hyderabad, Nizamabad, Aurangabad, Purnea, Bhagalpur, Gulbarga, Raichur, Bijapur, Bellary, Hospet, Cannanore, Calicut, Nanded, Bharauch, Saharanpur, Muzaffarnagar, Meerut, Bulandshahr, Aligarh, Firozabad, Etawah, Bareilly, Jaunpur and Bhagalpur. Muslims have a traditional association with these cities and are intrinsically linked to the social and economic life of these urban places. The Muslim population is also sizeable in many other cities and towns of India.

The Cultural Topography : As in other regions of the world, the Muslim population of India is divided into two sects, *viz.*, Shia and Sunni. The Sunnis are, however, in majority. In fact, the Shia population may not be constituting more than 10 per cent of the total Muslim population. Moreover, the Shias are mostly concentrated in a few pockets, such as Lucknow, Amroha, Allahabad, Kashmiri Gate area of Delhi, Rampur, Hyderabad and in the Kargil district of Jammu and Kashmir. But they are also found, although in small numbers, in almost every town and city of India. In such situations they generally prefer to live in a cluster. They are generally a well-knit community and their social and cultural activities are intrinsically linked to an *Imambarah,* which acts as the religio-cultural hub of the community.

In areas of their sizeable population they generally have separate mosques. Elsewhere, they offer prayers in the common mosques. They, however, follow a separate schedule of time for offering daily prayers. The Shias follow a separate theological tradition. In fact, there are striking differences between the Sunnis and the Shias insofar as the interpretation of certain aspects of the Islamic law is concerned. They are the followers of a lineage of *Imams* and *Mujtahids* descending down from Hazrat Ali and his two sons, Imam Hassan and Imam Hussain. These practices put them in a separate category both in terms of the interpretation of *Shariah* and the Quranic law as well as the Islamic customs.

The Sunnis, who constitute an overwhelming majority of the Muslim population of India, follow several schools of theological thought, notable among them being Hanafi, Hambali, Shaafai and Maliki. The Hanafis are, however, predominant. There are shades of differences among these schools in the interpretation of Islamic law as well as in Islamic practices. The founders of these schools—Imam Abu Hanifa, Imam Hanbal, Imam Shaafay and Imam Malik—held divergent views on many issues basic to Islamic theology. The interpretation of Islamic law *(Shariah)* has given rise to several traditions of Islamic jurisprudence. Another peculiarity of Islam in India is the rise of the two indigenous schools of theological thought which emerged at Deoband (in Saharanpur district of Uttar Pradesh) and at Bareilly.

These schools wield great influence on the Muslim masses who have been virtually divided into the Deobandis and the Barelwis. The *Dar-ul-Uloom* at Deoband, an internationally known institution of Islamic learning, often issues *fatwas* (religious decrees) on religious, legalistic, social, cultural and even political matters of general public concern. The divergent views held by these schools have created a wedge within the Sunni community. The Barelwi school which is more liberal admits Sufi tradition as the necessary ingredient of Islamic faith. The Deobandi school, on the other hand, is highly critical of any form of *mazar* worship.

Several groups, such as the Ismaiilis (followers of the Agha Khan), Bohras and Khojas, mainly found in the states of Gujarat and Maharashtra, are a category in themselves. The leaders of these religious sects, who wield great power, determine the course of life of their followers. Their say is supreme even in secular matters.

Places of Pilgrimage : The *dargahs* of the Sufis have emerged as places of pilgrimage par excellence. While there are several Sufi *silsilahs* in India, such as Chishtiya, Suharwardiya or Naqshbandiya, this description is restricted to the *dargahs* of the Chishtiya *silsilah* only. These *dargahs* have a hierarchy determined by their local, regional or national significance. Ajmer is at the apex of this hierarchy because the founder of the *silsilah*, Khwaja Moinuddin Chishti, is buried there. His *urs* (ceremonies related to the death anniversary) is of great national and international importance. Pilgrims assemble at his tomb for the annual *urs* ceremonies. In some years, the congregation may be as large as 5 lakh persons or even more. The pilgrims, however, continue to pour in throughout the year. They come from all over the country as well as the neighbouring countries of Bangladesh, Pakistan, Maldives and Mauritius.

The British Governor-General, Lord Curzon, once remarked: "There is a grave at Ajmer which rules over Asia." In hierarchical order, Ajmer is followed by the *durgah* of Bakhtiyar Kaki at Mehrauli in Delhi. The *durgah* of Sheikh Nizamuddin Auliya in Delhi is the next in hierarchy, which is followed by the *durgah* of Amir Khusrau, also in Delhi. These *dargahs* have an all-India importance. Then, there are other *dargahs* which have regional importance. Notable among them are those at Gulbarga (Karnataka), Kaliar Sharif (Uttar Pradesh), Kachhauchha Sharif (Uttar Pradesh), Nagaur (Rajasthan), Behsauri Sharif (Uttar Pradesh), Gorakhpur (Uttar Pradesh) and Charar-i-Sharif (Jammu and Kashmir).

It is interesting to note that while the Muslim *ulamas*, particularly of the Deoband school, have been opposed to the sufi practices, the Muslim masses are deeply involved in these practices. The *urs* ceremonies continue to be a distinguished feature of popular Islam. This is despite the fact that the *ulamas* consider the *urs* ceremonies as a heresy. This is one example where the *ulamas* advice is not heeded to by the Muslim masses. However, in many other matters, such as interpretation of marriage, and divorce laws, and the laws of inheritance, the *ulama* continue to hold their sway. They are recognized as the interpreters of the *Shariah*.

6

Geography and Language

Resources for Interaction

India is the home of a very large number of languages. In fact, so many languages and dialects are spoken in India that it is often described as a 'museum of languages'. The language diversity is by all means baffling. In popular parlance it is often described as 'linguistic pluralism'. But this may not be a correct description. The prevailing situation in the country is not pluralistic but that of a continuum. One dialect merges into the other almost imperceptibly; one language replaces the other gradually. Moreover, along the line of contact between two languages, there is a zone of transition in which people are bilingual. Thus languages do not exist in water-tight compartments. While linguistic pluralism is a state of mutual existence of several languages in a contiguous space, it does not preclude the possibility of inter-connections between one language and the other. In fact, these links have grown over millennia of shared history.

While linguistic pluralism continues to be a distinctive feature of the modern Indian state, it will be wrong to assume that there has been no interaction between the different groups. On the contrary, the give-and-take between the languages groups has been very common, often resulting in systematic borrowings from one language to the other. The cases of assimilation of one language into the other are also not uncommon.

Let us look at the nature of linguistic diversity observed in India today. According to the Linguistic Survey of India conducted by Sir George Abraham Grierson towards the end of the nineteenth century, there were 179 languages and as many as 544 dialects in the country. However, this number has to be taken with caution. It may even be misleading in the sense that dialects and languages were enumerated separately, although they were taxonomically part of the same language. Of the 179 languages as many as 116 were speech-forms of the Sino-Tibetan family, spoken by small tribal communities in the remote Himalayan and the northeastern parts of the country. Even the 1961 census recorded 187 languages. This was despite the fact that the census investigation was far more systematic and the classification was based on modern linguistic criteria. Much of this diversity may be understood properly in the light of some more statistics. For example, 94 out of the 187 languages were spoken by small populations of 10,000 persons or less. In the final analysis, about 97 per cent of the country's population was found affiliated with just 23 languages.

The diversity of languages and dialects is a reality and it is not the numerical strength of the speakers of a language which is important. The important fact is that there are people who claim a certain language as their mother tongue.

Another related development which contributed to linguistic diversity was the development of script. Different Indian languages were written in different scripts. This made learning of different languages a difficult exercise. However, with the growth of scripts, written languages have been successful in maintaining their record with the consequence that literary traditions have evolved. With the development of a script, oral communication is supplemented by a more powerful form of written communication. In the course of time some of the minor dialect and language groups have lost their identity as they have been assimilated into developed languages. It is a known fact that most of the languages still serve the purpose of oral communication only as speakers of these languages are still illiterate, or pre-literate.

It may be assumed that in the beginning various speech communities were confined to their own enclaves, more or less unaware of the existence of other language/dialect groups in the neighbourhood. Sometimes, the boundary between two dialects, or two languages, was knife-edged, as it was described by a hill-line or a river. Within the enclaves, these groups have been communicating through a common language, or dialect, for centuries. This became the basis of their identity. This traditional association with a language gave them a sense of belonging and thus inculcated in them a feeling of unity with the larger speech community. It may, however, be noted that inter-communication through a language or dialect is always limited in space. Individuals in their daily course of life have a limited reach. In situations where the communication is largely oral the sphere of communication is even smaller. Thus, with the passage of time, each speech community gets differentiated from other communities in the neighbourhood. This process leads to splitting of the spoken language into diverse dialects. The dialect formation is, however, within the same speech area. With the expansion of the speech territory more dialect groups emerge and the distance between them increases. Sometimes the outlying dialects are so isolated from the parent language that they acquire linguistic nuances of their own sufficient enough to be recognized as an independent branch.

A study of historical linguistics reveals that India has gone through all these phases of language development. The present linguistic map of India is naturally a product of these developments.

Different Tongues

The faculty of speech is by far the most distinctive human trait. Human beings use a system of language for communication which distinguishes them from the rest of the animal kingdom. It was through language that communication between the different members of a human group started in the early stages of social evolution. Language thus facilitated multiple forms of human cooperation. Eventually, a division of labour emerged, a prototype of which is unimaginable in the animal world.

However, there is no gainsaying the fact that animals also communicate with one another. They also produce vocal sounds although their sounds are simple. One can differentiate between warning calls, mating calls and those expressing anger or affection. This system of communication is simple as it lacks structure. A structured language was the invention of the human mind, and the most effective tool of communication. In this language words could be replaced easily to change the content of meaning.

In its basic characteristics a human language is essentially a signalling system in which a variety of vocal sounds are employed. These vocal sounds are produced by the peculiar constitution of the human speech organs. There is a combination of the different speech organs—the tongue, glottis, vocal cords and the palate—in producing vocal sounds which are essential elements in human articulation of language. It appears that in the beginning speakers of a language restricted their communication to relatively small number of vocal sounds out of the many sounds which human beings were capable of making. However, the number of vocal sounds varied from language to language which indicated variations in social evolution and the material conditions of existence. Most languages are satisfied with the use of twenty or thirty such sounds. But there are other languages which have as many as sixty sounds, or even more. There are others which have less than twenty sounds.

These sounds constitute the system of a language. As we know the purpose of a language is communication and a language sooner or later tends to become symbolic, more complex and expressive of abstract ideas. The beginnings of all languages were, however, simple.

There are very specific purposes for which the language is used. The main purpose is, of course, to express oneself, to convey one's feelings, sometimes to express a desire or pray for help. The human beings also communicate their ideas through body language either alone or in combination with vocal sounds or words. Even today, the body language continues to be a powerful means of expression. The exchange of ideas, feelings and calls for pray or help continue in the daily course of life of a human being. Thus, an intricate pattern of human cooperation and a feeling of togetherness is evolved. It is obvious that a language needs a group of people among whom communication continues through this language. This group of people who communicate in a certain language may be described as the 'speech community'. In the course of time, several speech communities are formed, each occupying a chunk of a geographically contiguous space.

Each language eventually expands over a territory, homogeneous in terms of its language structure—vocal sounds, words, sentences and conventionalized symbols. When a language is written in a script it tends to stabilize its distinguishing features and promotes communication over long distances between people.

The Genesis

Origins of language are shrouded in mystery. However, it is possible to reconstruct the bits of this history. It is generally agreed that in the history of social evolution language must have arisen with the discovery of the art of tool-making. Understandably, the early tool-making communities must have depended on cooperation between different members of the group on a highly organized basis. This would have been possible through the use of a language. Thus evolution of language must have progressed hand in hand with the evolution of material cultures. As the history of material cultures shows the change in techniques of tool-making was initially slow, but later on it picked up. Language also evolved with the same pace. Expressions became more and more complex with the passage of time. In fact, at every stage of evolution, there was a direct relationship between material culture and the language in use. Evidently progress in material cultures shows that the functions of brain were becoming more and more complex and with these changes language also became complex. The way languages evolved from vocal sounds to words and sentences revealed how they became symbolic as humans tended to express abstract rather than concrete ideas.

> *Language Families :* The process of change in a language often leads to divergent development. Imagine a language which is spoken only by the population of two small adjacent villages. In each village, the language will slowly change, but the changes will not be identical in both villages.... If communications between the two villages are bad, and members of one village seldom meet anybody from the other, then the rate of divergence may well be high. When a language is diverged into two forms like this, we say that it has two *dialects*.... Suppose now that the inhabitants of one of the villages pack up their belonging and migrate *en*

> *masse.* They go off to a distant country and under conditions quite different from their own home.... After a few hundred years the two dialects may have got so different that they are no longer mutually intelligible. We should now say that they were two different languages.... When two languages have evolved in this way from some single earlier language, we say that they are related (it is how the families of languages are formed).
> —Charles Barber, *The Story of Language*, 1964: *70-71*

It is obvious that in the course of evolution many languages were invented independently at different points of time in different regions of the world. They became further ramified as the social space within which inter-communication continued was always limited (Box). As a result, new groups were formed and new speech communities came into being. This is how the 'families of languages' developed. It is also understandable that the early languages were oral and writing became possible much later. In the beginning there was no need for maintaining a written record. When such a need arose writing was initially mostly pictorial. The discovery of the script in the history of development of languages must have taken a painstakingly long time during which picture/signs became conventionalized. Our knowledge of the early scripts is still incomplete. For example, the script of the Indus valley (Harappan) civilization continues to pose difficulties. We have not been able to decipher it simply because we are not familiar with the system of language in which communication was conducted by the Harappan people.

Socio-linguistic Strength

Despite the widely perceived linguistic diversity India's unity as a socio-linguistic area is quite impressive. Several linguists have analysed the basic elements of India as a socio-linguistic area. Describing language as an 'autonomous system',

Lachman M. Khubchandani recognized the major characteristics of the speech forms of modern India. Each region of the country is characterized by the plurality of cultures and languages "with a unique mosaic of verbal experience". In Khubchandani's view modern languages of India represent a striking example of the process of diffusion, grammatical as well as phonetic, over many contiguous areas. However, he considers linguistic plurality only as a superficial trait. "Indian masses through sustained interaction and common legacies have developed a common way to interpret, to share experiences, to think." What has emerged, is a kind of organic plurality, although the geographical distribution of speech communities suggests a kind of linguistic heterogeneity. Some of the basic elements of India's linguistic unity may be seen in the fuzzy nature of language boundaries, fluidity in language identity and complementarity of inter-group and intra-group communication. Khubchandani also emphasized the need of linking languages with the ecology of cultural regions described by him as *kshetras.* As a language area India is being put to mutually contradictory linguistic interpretations which confuse the issue. Perhaps a better understanding of the linguistic scene can be developed if the static account of the multiplicity of languages is replaced by a recognition of the elements of cultural regionalism. Similarly, the issue of linguistic homogeneity which has been argued by several linguists is fraught with complexities. In this context, one can cite the example of the states of the Indian Union carved out on the principle of linguistic homogeneity. The reality is that these states are not necessarily homogeneous in their language composition and cultural attributes.

In an earlier study, Khubchandani examined the evidence on plural languages and plural cultures of India. He dwelt upon the question of language in a plural society. The processes of language modernization and language promotion were also analyzed on the basis of a review of the language policies and planning in India. In this work, Khubchandani noted that people in certain regions of India displayed a certain degree of fluidity in their declaration of mother tongue. On this basis, he recognized two zones in which the country could be divided: a *fluid* zone and a *stable* zone. The fluid zone extended over the north-central region where Hindi, Urdu, Punjabi, Kashmiri and Dogri are spoken. The stable zone, on the other hand, incorporates western, southern and eastern regions. People in these regions did not reveal any fluidity in their mother tongue declaration.

Reference may also be made to the seminal work of Murray B. Emeneau who analyzed the characteristics of India as a language area. Tracing the history of development of the Indo-Aryan and the Dravidian languages he evaluated the shared experiences of the different speech communities. Emeneau defined linguistic area as "an area which includes languages belonging to more than one family but sharing traits in common which are found not to belong to other members of (at least) one of the families".

Languages : the Geographic Pattern : The geographic patterning of languages in the South Asian sub-continent can perhaps be understood in the context of the space relations the region had with other parts of Asia. As already pointed out, the sub-continent marks a southward projection of the Asian landmass into the Indian Ocean. The overland connections with West and Central Asia, Tibet, China and other regions of Southeast Asia helped the process of infiltration of linguistic influences into the South Asian region. This is evident from the fact that the languages spoken in the peripheral regions of South Asia, such as Baluchistan, Pak-Afghan borderlands, Kashmir, Gilgit, Hunza, Baltistan and Ladakh as well as the hilly parts of Himachal Pradesh and the regions in the Northeast have strong affinity with the languages spoken in the regions beyond the Hindu-Kush Himalayas. The remote Himalayan areas became the abode of Tibeto-Chinese languages. Similarly, the Northeastern region continued to receive influences from the neighbouring parts of Myanmar, Thailand and Indo-China. These regions are now the domain of the Tibeto-Chinese (Sino-Tibetan) or Tibeto-Burman languages. The people in the plains of North India from Sind to Assam acquired different branches of the Indo-European family of languages. The peninsular region continued to retain the Dravidian speech-forms even though the north was completely swayed over by the Indo-European languages. Between the Indo-European and the Dravidian one finds the Austric-speaking tribes nestled in the hills of the mid-Indian region.

The linguistic heterogeneity of India can perhaps be brought to some order when one realizes that these speeches really belong to four language families: Sino-Tibetan (Tibeto-Burman), Austro-Asiatic, Dravidian and Indo-European. In the course of usage over millennia of years these language families have found for themselves niches in the Indian social space in different parts of the sub-continent. Their geographical patterning throws some light on the routes through which these language families reached India. In fact, despite the vast heterogeneity, Indian languages experienced parallel trends in linguistic and literary development during the long phases of shared history. This has made India 'a composite region' in terms of linguistic attributes (Table).

Broad Classification of Modern Indian Languages

Family	*Sub-family*	*Branch/Group*	*Speech Area*
Austric			
(Nishada)	Austro-Asiatic	Mon-Khmer	Meghalaya, Nicobar Islands Munda West Bengal, Bihar, Orissa, Assam, Madhya Pradesh, Maharashtra
	Austro-nesian		Outside India
Dravidian *(Dravida)*		South Dravidian Kerala	Tamil Nadu, Karnataka,

Contd...

Family	*Sub-family*	*Branch/Group*	*Speech Area*
		Central Dravidian	Andhra Pradesh, Madhya Pradesh, Orissa, Maharashtra
		North Dravidian	Bihar, Orissa, West Bengal, Madhya Pradesh
Sino-Tibetan	Tibeto-	Tibeto-	Jammu and Kashmir,
(Kirata)	Myanmari	Himalayan	Himachal Pradesh, Sikkim,
		North Assam	Arunachal Pradesh
		Assam-Myanmari	Assam Nagaland, Manipur, Mizoram, Tripura, Meghalaya
	Siamese-Chinese		Outside India
Indo-European			
(Arya)	Indo-Aryan	Iranian	Outside India
		Dardic	Jammu and Kashmir
		Indo-Aryan	Jammu and Kashmir, Punjab, Himachal Pradesh, Uttar Pradesh, Rajasthan, Haryana, Madhya Pradesh, Bihar, Orissa, West Bengal, Assam, Gujarat, Maharashtra, Goa

Affiliation and Diffusion of Language : The history of Indian languages is not easy to reconstruct. As an overview of the processes of peopling of India shows, Negroids were the first people to arrive. However, we do not exactly know about their language affiliation. The subsequent waves of migrations were so strong that the Negroids lost their identity completely, leaving behind little traces of either their racial or linguistic past.

The story of the four families of languages may be briefly recapitulated here, although it is not easy to establish the chronological sequence in which the speakers of the Austric, Sino-Tibetan and the Dravidian languages came to India. It is almost certain that these families were already there at the time of the advent of the Indo-Aryan. This is, however, an established fact that the Sino-Tibetan speech communities were Mongoloids racially. The original Sino-Tibetan, the parent of the early Chinese, is supposed to have developed somewhere in western China around 400 B.C. It is also believed that the diffusion of this language eventually affected the regions lying to the south and the southwest of China-Tibet, Ladakh, northeastern India, Myanmar and Thailand. Perhaps, the Vedic Aryans were familiar with this group. They described the Tibeto-Burman-speaking Mongoloids of the Brahmaputra valley and the adjoining regions as *Kiratas.*

The speakers of the Kirata family of languages are distributed all along the Himalayan axis from Baltistan and Ladakh to Arunachal Pradesh. They occupy the regions surrounding the Brahmaputra valley in the northeast from Nagaland to Tripura and Meghalaya. There are striking differences between the languages of the Kirata family distributed over such a vast geographical area. The speakers of the Tibeto-Himalayan branch of the Kirata languages occupy the Himalayan regions from Baltistan to Sikkim and beyond to Arunachal Pradesh.

The Bhotia group consists of the Balti, Ladakhi, Lahauli, Sherpa and the Sikkim Bhotia dialects. Linguists also recognize a Himalayan group consisting of Lahuli of Chamba, Kanauri and Lepcha which is distinguishable on the basis of certain linguistic traits. In the east there is a North-Assam branch including the dialects of Arunachal Pradesh, such as Miri and Mishing. In other parts of the northeast the languages belong to the Assam-Burmese branch and are divided into Bodo, Naga, Kachin and Kuki-Chin groups.

The speakers of the Kirata languages came to India in different streams at different points of time. Understandably, the groups in the northwest were unrelated to the groups in the north-east. Similarly, the Kachin and the Kuki-Chin groups followed separate routes of migration. This is why there is a vast variety of dialects within the Kirata family and the roots of linguistic heterogeneity go far beyond the Indian borders into the neighbouring parts of Tibet, Myanmar and Indo-China.

Anthropologists as well as linguists believe that the Austric-speaking groups came earlier than the Dravidian-speaking communities. The Austric speech communities were already there in the mid-Indian region before the advent of the Dravidian. The present geography of the Austric dialect groups holds some clues to the historical processes of their diffusion into India. Generally, the Austric family of languages is recognized as consisting of a Mon-Khmer and a Munda branch. The Mon-Khmer speakers belong to two separate groups, *viz.*,

Khasi and the Nicobarese, both separated by a distance of more than 1,500 kilometres which spans over an expanse of the Bay of Bengal. There is no clarity among the scholars about the routes taken by the speakers of the Mon-Khmer dialects. The Khasi speakers themselves are surrounded by other Kirata and Arya dialects in the Meghalayan plateau.

The advent of the Dravidian in India is generally associated with a branch of the Mediterranean racial stock which was already there in India before the rise of the Indus valley civilization. In fact, archaeologists believe that they were the builders of the Harappan civilization along with the Proto-Austroloids.

The Dravidian speech communities were found over most of the northern and the northwestern region of India before the advent of the Indo-Aryan. However, following the rise of the Indo-Aryan in north-western India, a linguistic change came and the Dravidian-speaking area shrank in its geographical extent. The present distribution of the Dravidian dialects in different parts of North India, such as Baluchistan, Chhotanagpur plateau and eastern Madhya Pradesh, where Baruhi, Kurukh-Oraon and the Gondi are spoken respectively, suggest the earlier stage of distribution of this family of languages. In fact, Gondi is spoken in many parts of Central India from Madhya Pradesh and Maharashtra to Orissa and Andhra Pradesh. While Dravidian speech forms were in use for many centuries in the pre-Christian era, the literary development in the Dravidian speech community could take place only in the first few centuries after Christ. It is believed that the old Tamil, old Kannada and the old Telugu had already come into being by 1000 A.D. Malayalam acquired its form a little later.

With the Vedic Sanskrit, a branch of the Indo-European, the Indo-Aryan established itself in north-western India. It had definite relations with the different Indo-European languages, such as Persian, Armenian, Greek, French, Spanish, German and English. An early form of Indo-European seems to have genetic relations with the Hittite speech of Asia minor. The linguists have recognized a primitive form of Indo-European in its earlier stage of development. They called it Indo-Hittite. A branch of the Indo-European which had already established itself in Mesopotamia came to be described by the linguists as Indo-Iranian. It is this Indo-Iranian branch which spread over Iran and the northwestern regions of India by the middle of the second millennium B.C.

Among the different families of languages spoken in India the Indo-European seems to be the last to arrive. The advent of Indo-European in the South Asian sub-continent brought about a major change in the linguistic affinity of the people of northern India. The form of Indo-European which was spoken in India, came to be known as Indo-Iranian or Indo-Aryan. Its advent in India is seen with the rise of the Vedic Sanskrit. However, the old Sanskrit changed into Prakrit and several speech forms developed in different parts of northern and western India. The region lying between Saraswati and Ganga, encompassing the upper Ganga-Yamuna doab and adjoining parts of Haryana, to the west of the Yamuna, became the stage for the transformation of classical Sanskrit into a Prakrit form. From this early stage of development of Prakrit came the different Indo-Aryan vernaculars which are now spoken in north-western, north-central, central and eastern parts of India.

The Suraseni emerged in the core region of the midland (Madhyadesa of the *Puranas)* as the popular language. Its core area extended over western Uttar Pradesh and the adjoining parts of Haryana. A developed form of this parent language is described by the linguists as Western Hindi. Around the core region of Suraseni other speech forms developed on the west, south and east. These languages formed an outer band around the core language.

On the west and the northwest lay the Punjabi and the Pahari dialects. Rajasthani and Gujarati emerged on the southwest. On the east, a form of language, now known as Eastern Hindi, emerged in Kosali (Awadhi). Linguists believe that these outer dialects were all more closely related to each other than any one of them was to the language of the midland. "In fact, at an early period of the linguistic history of India, there must have been two sets of Indo-Aryan dialects—one the language of the midland and the other the group of dialects forming the outer band."

This first stage was followed by a subsequent phase of expansion. As the population of the midland region increased expansion became a necessity. Thus, on the periphery of the languages of the outer band developed new speech forms which were by and large not related to the language of the midland. For example, while Punjabi was closely related to the language of the upper doab it got transformed into Lahnda in southwestern Punjab. This language had little relationship with the language of the midland. With increasing distance changes became quite pronounced.

The geographical distribution of the Indo-Aryan languages may be briefly summarized here as follows: The midland language occupies the Ganga-Yamuna doab and the regions to its north and south. This core region is encircled by different speech forms in eastern Punjab, Rajasthan and Gujarat. Further beyond in the west and the northwest, there is a band of outer languages—Kashmiri, Sindhi, Lahnda and Kohistani. The languages of this band may be described as constituting the northwestern group of the outer languages.

On the southern periphery lies the Marathi. In the intermediate band are situated languages, such as Awadhi, Bagheli and Chhattisgarhi. On the eastern periphery lie the three dialects of Bihari, *viz.*, Bhojpuri, Maithili and Maghadi. The Bihari is surrounded by Oriya in the southeast and Bengali in the east. The languages of the eastern branch of the Indo-Aryan extend further in the east where Assamese occupies the Brahmaputra valley.

Linguists believe that the development of the Indo-Aryan languages completed itself through several phases. The Prakrits developed into two stages: Primary Prakrits and Secondary Prakrits. The *Primary Prakrits* whiqh were the first to evolve out of the classical Sanskrit were synthetic languages with a complicated grammar. In the course of time they 'decayed' into *Secondary Prakrits.*

"Here we find the languages still synthetic, but diphthongs and harsh combinations are eschewed, till in the latest developments we find a condition of almost absolute fluidity, each language becoming an emasculated collection of vowels hanging for support on an occasional consonant. This weakness brought its own nemesis and from, say 1000 A.D., we find in existence the series of modern Indo-Aryan vernaculars, or, as they may be called *Tertiary Prakrits.*"

The last stage of development of the Prakrits is known as literary *Apabrahmsa*. It is supposed that the modern vernaculars are the direct children of these Apabrahmsas. The sequence of change was like this. The Suraseni Apabrahmsa was the parent of Western Hindi and Punjabi. Closely connected with it were Avanti, the parent of Rajasthani, and Gaurjari the parent of Gujarati. The other intermediate language —Kosali (Eastern Hindi)—sprang from Ardha-Magadhi Apabrahmsa. The chronological sequence may be roughly reconstructed here (Table).

Stages in the Development of Indo-Aryan Languages

Period	*Language*	*Prakrits*
1500-600 B.C.	Vedic Sanskrit	
600-200 B.C.	Asokan Prakrits	Pali, Eastern/ Western
200 B.C-200 A.D.	Transitional Middle Indo-Aryan	
200-600 A.D.	Middle Indo-Aryan	Primary/ Secondary Prakrits
600-1000 A.D.	Late Middle Indo-Aryan	Apabrahmsas
Beyond 1000 A.D.	New Indo-Aryan	

These different stages through which the Indo-Aryan languages passed.

The Grouping

In a country where so many languages/dialects are spoken, and many of them are used for oral communication only, linguistic classification may not be an easy exercise. The scientific study of Indian languages, their grammar, phonetics and vocabulary which goes back to the nineteenth century, is still ridden with problems. For one thing, linguists are still unsure of genetic relationships between one language group and the other. Their knowledge of some of the minor languages is pathetically inadequate. This leaves the problem of classification always open to revision. A second set of problems arises from the recognition of the major languages and their specification in the Eighth Schedule of the Indian Constitution. There were political compulsions under which some languages were given this special status. The Eighth Schedule mentions eighteen languages; twelve of them have their own territory where they receive maximum state patronage and seem to have great potential for development.

The Eighth Schedule also includes languages, such as Sanskrit, Sindhi, Nepali and Urdu. The first three do not have a speech territory as such. The speakers of Urdu, Sindhi and Nepali are distributed across several states. Minor speech communities, such as Manipuri (with a population of 1.27 million) and Konkani (with a population of 1.72 million) have also been given the status of scheduled languages. The anomaly in this approach is that while some of the minority speech groups have found a place in the Eighth Schedule, major Austric languages, such as Santali (speakers 5.22 million), Bhili (speakers 5.57 million) or Gondi (speakers 2.12 million), have been completely ignored. The official language policy leaves the issue amenable to political manipulation. Like the Austric languages all small Tibeto-Burman languages have also been excluded from the Eighth Schedule. Hindi, which has the largest number of speakers, is an aggregate of at least fifty different dialect groups. There are at least seven states which recognize Hindi as their official language.

Of the four language families (Tibeto-Burman, Austro-Asiatic, Dravidian and the Indo-Aryan branch of the Indo-European) the most diverse is the Tibeto-Burman as their speakers communicate in 70-80 different languages. Next is the Indo-Aryan with 19 languages grouped under it. The Dravidian family incorporates 17 different speech communities; while the Austro-Asiatic has 14 languages. However, this analysis is incomplete because of the fact that the 1991 census used an eligibility condition for recognizing a language as a mother tongue if it had more than 10,000 speakers at the all-India level. This resulted in the exclusion of 0.56 million speakers of different minor languages from finding a reference in the census records. There is, therefore, no recognition of these small speech communities.

Due to operational difficulties, the 1991 census could not be conducted in the state of Jammu and Kashmir. This resulted in the exclusion of the Dard group of languages (Dardi, Shina, Kohistani and Kashmiri) from the census count. The compilation of data for the 18 scheduled languages also contributed to the multiple problems of classification. For example, minor speech communities, such as Chakma and Hajong, were clubbed together with Bengali. Secondly, Yerava and Yerukala were amalgamated with Malayalam and Tamil respectively. Such examples are many. It is obvious that the census preferred to ignore the minor dialect groups.

The story of Hindi is equally interesting. For example, 50 dialect groups have been grouped under Hindi with the result that there is no scope for an analytical study of the geographical spread of these dialect groups. In the course of time many speakers of these dialects have tended to declare Hindi as their mother tongue without making reference to the dialect they use. This is evident from the fact that some 233 million speakers declared Hindi as their mother tongue. Gradually, the stage is not far when many of the distinguished dialect groups, such as Braj Bhasha, Awadhi, Bhojpuri, Magadhi, Maithili and Marwari, will loose their identity at least in the census records. This withdrawal of patronage by the speakers of these dialects is a symptom of their eventual decline, if not death.

A broad classification scheme of the four language families is given in Table (for detailed classificatory schemes, see Tables).

Classification of Austric Languages

Branch	*Language*	*Speakers, 1991*	*Per cent of All Speakers*
Mon Khmer	Khasi	912,283	9.64
	Nicobarese	26,261	0.28
Munda	Santhali	5,216,325	55.13
	Munda	413,894	4.37
	Mundari	861,378	9.10
	Bhumij	45,302	0.48
	Birhor Koda/Kora	28,200	0.30
	Ho	949,216	10.03
	Asuri	NA	NA
	Brijia	NA	NA
	Korku	466,073	4.93
	Kharia	225,556	2.38
	Juang	16,858	0.18
	Savara	273,168	2.89
	Gadaba	28,158	0.30
	Kharwari		
	Total Speakers	9,462,672	100.00

Source: Census of India, 1991, Series-1, India, Paper 1 of 1997.

Classification of Tibeto-Burman Languages

Branch Group	*Language*	*Speakers, 1991*	*Per cent of All Speakers*
Tibeto-Himalayan Bhotia	Tibetan	69,416	0.86
	Balti	NA	NA
	Ladakhi	NA	NA
	Lahauli	22,027	0.27
	Sherpa	16,105	0.20
	Sikkim Bhotia	55,483	0.69
Himalayan	Kinnauri	61,794	0.76
	Lepcha	39,342	0.49
	Lahauli	—	—

Contd...

Branch Group	*Language*	*Speakers, 1991*	*Per cent of All Speakers*
North-Assam NEFA	Aka	NA	NA
	Dafla	NA	NA
	Abor/Adi	NA	NA
	Adi	158,409	1.96
	Miri**	390,583	4.83
	Mishmi	29,000	0.36
	Mishing	NA	NA
Assam-Burmese Bodo	Bodo	1,221,881	15.10
	Kachari	NA	NA
	Lalung	33,746	0.42
	Dimasa	88,543	1.09
	Garo	675,642	8.35
	Koch	26,179	0.32
	Rabha	139,365	1.72
	Mikir/Karbi	366,229	4.53
	Tripuri	694,940	8.59
Naga	Angami	97,631	1.21
	Sema	166,157	2.05
	Rengma	37,521	0.46
	Khezha	13,004	0.16
	Ao	172,449	2.13
	Lotha	85,802	1.06
	Mao	77,810	0.96
	Tangkhul	101,841	1.26
	Konyak	137,722	1.70
	Pochury	11,231	0.14
	Phom	65,350	0.81
	Yin Chungre	47,227	0.58
	Khiemnungan	23,544	0.29

Contd...

Branch Group	*Language*	*Speakers, 1991*	*Per cent of All Speakers*
Kuki-Chin	Manipuri/Meitei	1,270,216	15.70
	Thado	107,992	1.33
	Paite	49,237	0.61
	Lushai	538,842	6.66
	Rangkhul	NA	NA
	Halam	29,322	0.36
	Hmar	65,204	0.81
	Total Speakers	8,092,940	100.00

The census did not count Lahauli of Chamba separately, although it was other than the Lahauli mentioned under Bhotia group.

**includes Mishing.

*** includes Kachari. Although total speakers of Kachari were separately shown as 11588.

Total speakers of all languages of Tibeto-Burman family.

Source: Census of India, 1991, Series-1, India, Paper 1 of 1997.

Classification of Dravidian Languages

Group	*Language*	*Speakers, 1991*	*Per cent of A11 Speakers*
South Dravidian	Tamil	52,886,931	27.99
	Yerukala	63,133	0.03
	Malayalam	30,325,637	16.05
	Yerava	17,295	0.01
	Kannada	32,753,676	17.34
	Kuruba	NA	NA

Contd...

Group	*Language*	*Speakers, 1991*	*Per cent of A11 Speakers*
Coorgi/Kodaga		97,011	0.05
	Tulu	1,550,334	0.82
	Toda	NA	NA
	Kota	NA	NA
	Koda/Kota	27,703	0.01
Central Dravidian	Telugu	65,900,723	34.88
	Vadaga	NA	NA
	Kui	641,600	0.34
	Kolami	98,281	0.05
	Gondi	2,124,852	1.12
	Parji	44,001	0.02
	Koya	270,994	0.14
	Khond/Kondh	220,783	0.12
	Konda	17,864	0.01
North Dravidian	Kurukh-Oraon	1,426,618	0.76
	Malto	108,148	0.06
	Total Speakers	188,945,126	100.00

Source: Census of India, 1991, Series-1, India, Paper 1 of 1997.

Classification of Indo-European Languages

Branch	*Group*	*Language*	*Speakers, 1991*	*Per cent of All Speakers*
Dardic	Dard	Dardi	NA	NA
		Shina	NA	NA
		Kohistani	NA	NA
		Kashmiri	56,693	0.01
Indo- Aryan	Northwestern	Lahnda	27,386	0.00
		Sindhi/ Kachchhi***	2,122,848	0.34

Contd...

Branch	*Group*	*Language*	*Speakers, 1991*	*Per cent of All Speakers*
	Southern	Marathi	62,421,442	9.89
		Konkani	1,723,264	0.27
	Eastern	Oriya	27,586,476	4.37
		Bihari		
		a. Maithili	7,766,597	1.23
		b. Magadhi	10,566,842	1.67
		c. Bhojpuri	23,102,050	3.66
		Bengali	66,552,894	10.54
		Assamese	12,962,721	2.05
	East-central	Awadhi	481,316	0.08
		Bagheli	1,387,160	0.22
		Chhattisgarhi	10,595,199	1.68
	Central	Hindi (Western Hindi)		
		a. Bangaru		
		b. Braj Bhaka	85,230	0.01
		c. Bundeli	1,657,473	0.26
		d. Kanauji		
		Urdu	43,406,932	6.88
		Punjabi/East Punjabi	23,085,063	3.66
		Rajasthani	13,328,581	2.11
		a. Marwari	4,673,276	0.74
		b. Mewari	2,114,622	0.33
		c. Mewati	102,916	0.02
		d. Kurmali Thar	236,856	0.04
		e. Bagri Rajasthani	593,730	0.09
		f. Banjari	887,632	0.14
	Northern	Eastern Pahari/Nepali	2,075,746	0.33
		Central Pahari		
		a. Kumauni	1,717,191	0.27
		b. Garhwali	1,872,578	0.30
		Western Pahari		

Contd...

Branch	*Group*	*Language*	*Speakers, 1991*	*Per cent of All Speakers*
		a. Jaunsari	96,995	0.02
		b. Sirmauri	370,558	0.06
		c. Kului	152,442	0.02
		d. Mandeali	440,421	0.07
		e. Ghameali	63,408	0.01
		f. Bharmauri/ Gaddi	18,919	0.00
		g. Churahi	45,107	0.01
		Total Speakers	631,273,191	100.00

*The numerical strength of the Kashmiri speakers shown here is incomplete as 1991 census was not conducted in Jammu and Kashmir. The figure includes 22,398 speakers of Siraji.

**Includes 20,589 speakers of Multani.

***Sindhi 1,551,384; Kachchhi 566,199.

+ A total strength of 13328,581 has been shown separately under Rajasthani.

+ + A total strength of 2179,832 is shown under Pahari; eastern, western or central categories not indicated.

Source: Census of India, 1991, Series-1, India, Paper 1 of 1997.

Numerical Strength : Of the four language families Indo-European has by far the largest strength of speakers. In fact, three-fourths of the country's population claimed one or the other language of the Indo-European family as their mother tongue. The Dravidian family comes next with 22.5 per cent of the total population of the country claiming affinity to it. The speakers of the other two families—Austro-Asiatic and the Tibeto-Burman—consist of small groups. Their overall proportionate share is low: 1.13 and 0.97 per cent respectively. As already indicated, Austro-Asiatic languages are spoken by a host of tribal groups. This is also true for most of the Tibeto-Burman languages.

Among the languages of the Austro-Asiatic family, Santali is the most outstanding speech community, the numerical strength of its speakers being as high as 5.2 million. Other languages of the Munda branch, such as Ho, or of the Mon Khmer branch, such as Khasi, have a numerical strength of less than one million speakers each. Santali speakers account for 55 per cent of the entire strength of the Austro-Asiatic family. Speakers of the Ho, Khasi and Mundari languages account for 10, 9.6 and 9.1 per cent respectively of all Austric speakers. There are many language groups within the Austro-Asiatic family whose numerical strength is insignificant. Reference may be made to Bhumij, Nicobarese, Gadaba and Juang. However, their declining numerical strength shows that conditions are not favourable for their growth. As indicated earlier, the 1991 census adopted a policy of excluding all languages from the census count whose speakers numbered less than 10,000 persons at the all-India level at the time of census enumeration. This policy was by and large negative to the interests of the tribal languages.

As is evident from Tables, there are striking differences in the numerical strength of the languages of the Tibeto-Burman family. The major speech communities include Manipuri/ Meithei (1.27 million), Bodo (1.22 million), Tripuri (0.69 million), Garo (0.67 million), Lushai (0.54 million) and Miri/Mishing (0.39 million). The Manipuri and Bodo groups together account for one-third of the total strength of Tibeto-Burman speakers.

As is generally known, the major Dravidian speech communities consist of Telugu, Tamil, Kannada and Malayalam. They have been ranked here in descending order of the strength of their speakers. The Dravidian family also includes minor groups, such as Gondi (2.12 million), Tulu (1.55 million), Kurukh-Oraon (1.42 million), Kui (0.64 million), Koya

(0.27 million) and Khond (0.22 million). Many of them are tribal dialects and have an imminent risk of extinction. Gondi, for example, presents a case of language loss, as its speakers are getting assimilated into the regional languages of the state of their habitation. The same is true for other tribal dialects, unless otherwise they have come under the cover of state protection.

In terms of numerical strength of speakers Hindi is the foremost among the Indo-European languages. With 337.27 million speakers who claimed Hindi, or its different dialects, as their mother tongue, Hindi has no comparison with other languages of the family. Bengali, Marathi and Urdu which follow in the same order have a numerical strength ranging between 43 and 69 million. Bengali and Marathi account for about 10 per cent each of the total strength of speakers of the Indo-European family. Among the dialects of Hindi, Bhojpuri was claimed by 23.1 million speakers. One can compare Bhojpuri with Assamese (total speakers: 12.96 million) and Punjabi (total speakers: 23.08 million). The other dialects of Hindi, such as Maithili, Magadhi, Awadhi, Braj Bhasha, Marwari or Chhattisgarhi figure poorly. In fact, the dialect speakers tend to declare Hindi as their mother tongue. The strength of those who declared these dialects as their mother tongue seems to be diminishing with successive censuses. The progress of the languages of the Dard group, namely, Shina, Kohistani and Kashmiri, cannot be monitored since 1991 census was not conducted in Jammu and Kashmir.

As many as 13 of the Indo-European languages have been listed in the Eighth Schedule of the Indian Constitution. They are: Kashmiri (Dard group), Sindhi (northwestern group), Hindi, Urdu, Punjabi, Nepali, Gujarati (eastern, east-central, central and northern groups), Bengali, Assamese and Oriya (eastern group) and Marathi and Konkani (southern group). While Konkani, with a total strength of 1.76 million speakers, is mentioned in the Eighth Schedule, Santali finds no place, although its speakers numbered at 5.22 million at the 1991 census.

World of Tongues

A generalized study of the domains of various languages spoken in India may be helpful in understanding the historical processes leading to their geographic spread and concentration. It may also be helpful in identifying the basic elements of India's linguistic geography. It may be worthwhile to recapitulate the historical processes that led to the evolution of language regions in India (Box). It is understood that the Indo-Aryan was the last to arrive. It was preceded by Dravidian, Sino-Tibetan and Austric. However, there is no clarity about the chronological sequence in which the different families came to affect the situation in India. This question has been partly answered by the linguists. Which came first? Austro-Asiatic, Sino-Tibetan or Dravidian? The Vedic Aryans had the knowledge of the Tibeto-Burman-speaking Mongoloids whom they described as the *Kiratas*. *Yajurveda* and *Atharvaveda* as well as Mahabharata and Manu Samhita also mentioned the *Kiratas*.

Growing Period

Among the four language groups which were established in India in very early times, the Aryan speech family was the last to come. It was preceded by the Dravidian, Sino-Tibetan and Austric. In the evolution of Indian people and Indian culture, there has been an intermixture of races, languages and the cultural *milieus*. Although the Austric and Sino-Tibetan languages are now confined to small populations, they have had their share in developing and modifying the other languages. The Dravidian and the Aryan speeches have, of

> course, been the most important of all; after the Aryan, the Dravidian was the first to develop literature. The other groups had no written literature until long after. There has been a good measure of interaction among these languages. The Aryan has been profoundly modified by the Dravidian, and vice versa; and that applies to other languages as well. This kind of linguistic interaction has given rise, after 3,000 years of free play, to a Common Indian Type for the modern languages of all the four families of speech, through the evolution of a certain amount of similarity in phonetic structure, morphology, and above all, syntex and vocabulary... Indian languages show... important resemblances. That is because they share in a common or pan-Indian character evolved through the racial and linguistic intermixture which has been at work since the beginning of history.
>
> —S.K. Chatterji and S.M. Katre, *The Gazetteer of India*, Vol. I, 1975: 373-374

Austro-Asiatic Languages : The domain of the Austro-Asiatic languages lies in the mid-Indian region and extends from Maharashtra to West Bengal. The two outliers of this domain—Khasi and Nicobarese—have their enclaves in Meghalaya and Nicobar Islands respectively. The two pockets are separated by a vast expanse of the sea. Santali is the foremost among the Munda languages. The Santali speakers are mainly concentrated in Bihar, West Bengal and Orissa. About one-half of them live in Bihar, 35 per cent in West Bengal and 13 per cent in Orissa. The Santhals living in Assam, numbering 135,000, also declared Santali as their mother tongue at the 1991 census. Another significant language of the Munda branch is Munda/ Mundari. Of the 1.27 million speakers of Mundari, 54 per cent live in Bihar, 31 per cent in Orissa and only 6 per cent in West Bengal. The domain of the Ho language lies in Bihar and

Orissa. Two-thirds of all Ho speakers are confined to Bihar and the remaining one-third to Orissa. The territories of the Kharia, Korku and Savara languages extend over Bihar, Orissa and Madhya Pradesh. However, the Savara speakers are mostly confined to Orissa.

Tibeto-Burman Languages : The territory of the Tibeto-Burman languages is by and large conterminous with the Himalayas and extends from Baltistan and Ladakh in Jammu and Kashmir to Arunachal Pradesh. It extends further to encompass other northeastern states. The Bhotia and the Himalayan groups of the Tibeto-Burman family are confined to Jammu and Kashmir, Himachal Pradesh, hilly Uttar Pradesh and Sikkim. The Tibetan speakers, however, have a wider spread as they are distributed over many states in India. The Tibetans are of course in exile in India and live in camps and colonies especially created for them in several states. Notable among the languages of the North Assam branch of the Tibeto-Burman family are Miri/Mishing and Adi. More than 97 per cent of the Adi speakers are confined to Arunachal Pradesh. The Miri/Mishing speakers, on the other hand, are confined to Assam.

Of the languages of the Bodo group, Bodo is largely specific to Assam where 97 per cent of its speakers live. The domains of the Garo and Tripuri lie in Meghalaya and Tripura respectively. About 80 per cent of the Garo speakers are confined to Meghalaya whereas 17 per cent of them are based in Assam. About 93 per cent of the Tripuri speakers are confined to Tripura, although in recent years a section of their population has also moved out to Mizoram and Assam. The Bodo group also includes Karabi/Mikir and Rabha dialects. Their speakers are mostly concentrated in Assam and Meghalaya. However, a small proportion of Rabha speakers is also found in the northern districts of West Bengal. Likewise, the Koch is confined to Meghalaya and Assam, Dimasa to Assam and Nagaland, and Lalung is specific to Assam alone.

Most of the speech territory of the Naga group of languages is shared between Nagaland and Manipur. While Ao, Angami, Lotha, Pochury, Phom, Yimchingure and Khiemnungan are exclusive to Nagaland, Kabui and Tangkhul are specific to Manipur. On the other hand, Khezha and Mao are spoken both in Manipur and Nagaland. However, a small proportion of their speakers is also located in Assam. The Kuki-Chin languages are confined to the states of Manipur and Mizoram. Manipuri (including Meitei) has its domain in the central valley of Manipur where 87 per cent of its speakers live. A section of the Manipuri population (about 10 per cent) has also moved out to Assam. Manipuri speakers were also enumerated in Tripura, Nagaland and other parts of the northeast, although in small numbers. Lushai is confined to Mizoram.

Manipur presents a case of linguistic plurality. In fact, the state is the home of many speech communities belonging to both the Naga and Kuki-Chin groups. Notable among these dialects are Thado, Paite, Halam, Hmar, Kabui, Tangkhul, Gangte, Khezha, Kom, Kuki, Liangmei, Lushai, Mao, Maram, Maring, Vaiphei, Zeliang, Zemi and Zau. Lakher has its domain in Mizoram only. Migration in recent years has taken the Kuki speakers to other parts of the northeast, such as Assam, Nagaland and Tripura.

Manipur may be chosen as an example to illustrate the territoriality of minor language groups in a contiguous geographical space. The people of Manipur exhibit a complex pattern of ethnic diversity, where each ethnic or dialect group tends to concentrate in a monolithic world of its own. Broadly speaking, the population of Manipur consists of two different groups: (a) Palaeo-Mongoloids consisting of (i) the Meiteis, and (ii) the hill tribes; and (b) immigrants mostly consisting of Palae-Mediterraneans further sub-divided into (i) the Pangals (Muslim settlers), and (ii) the Mayangs or Kols. Each of these groups can be further sub-divided on the basis of language/

dialect and racial attributes. While the Meiteis, Pangals and Mayangs are plain-dwellers, the tribes, such as Kuki-Chins and Nagas are hill-dwellers. The ethnolingual situation in Manipur suggests that geographical factors have promoted the emergence of homogeneous dialect territories. Each dialect is confined to a pocket where people communicate in a given dialect. These monolithic dialect territories are contiguous. This linguistic plurality has survived the onslaught of time.

In his doctoral thesis Hemkhothang Lunghdim examined the patterns of communication in the multi-speech area of Manipur. The presence of as many as 29 major speech communities in Manipur has contributed to a type of ethno-centrism for the survival of speeches or *Patois* coupled with other socio-cultural differences. Ethnic groups have often indulged in competition resulting in inter-ethnic conflicts. In any case ethnic groups strive for the preservation of their dialects even if it results in diminishing interaction between different dialect groups. Over time linguistic plurality has resulted in bilingualism or multilingualism. Many tribes have adopted elements of Meitei-Lon for mutual inter-communication (Box).

> ***Language Plurality in Manipur :*** Within the state, this bilingualism or multilingualism appears to have a certain detectable trend or pattern among the different populations. First, it is noticed that the tribals are usually multilingual. For instance, Kuki-Chin tribals, even before coming into contact with Meitei, were already bi-or multilingual, as necessitated by their environment, although in their village or home they use one speech. Secondly, bilingualism or multilingualism in the case of Meiteis or 'Mayangs' has not been in the tribal dialects. A Meitei need not be bilingual so long as he/she stays in Manipur. Bilingualism to Meiteis would generally mean fluency in other Indian

> languages, and in very exceptional cases it may include tribal dialects. This situation is true for all plain settlers. Thirdly, the pattern is slightly different for the 'Mayangs', who are mainly merchants, or consist of army personnel. To them, the direction of bilingualism or multilingualism could be towards all the ethnic dialects. However, even to them knowledge of Meitei-Lon is essential and always beneficial.
>
> —Hemkhothang Lhunghdim in Itagi (ed.), *Spatial Aspects of Language*, 1994: 63

Dravidian Languages : As is generally known, the four southern states of Andhra Pradesh, Karnataka, Tamil Nadu and Kerala are the home of the major Dravidian languages, *viz.*, Telugu, Kannada, Tamil and Malayalam. The speakers of these languages have also moved out to other states, particularly the neighbouring states of the south in the recent past. The geographical spread of these languages is evident from Table.

Distribution of Major Dravidian Languages by States, 1991

Language	*States*
Telugu	Andhra Pradesh (83.40), Tamil Nadu (6.02), Karnataka (5.04)
Kannada	Karnataka (90.94), Tamil Nadu (3.69), Maharashtra (3.24), Andhra Pradesh (1.59)
Tamil	Tamil Nadu (91.38), Karnataka (3.26), Andhra Pradesh (1.42), Pondicherry (1.36)
Malayalam	Kerala (91.79), Tamil Nadu (2.29), Karnataka (2.65), Maharashtra (1.33)

Note: Figures in parentheses show the proportion of speakers of respective languages. *Source:* Table compiled by the author.

Evidently, these languages display a high degree of concentration in their home states. The highest degree of concentration is seen in the case of Malayalam, followed by Tamil. On the other hand, the lowest degree of concentration is revealed in the case of Telugu.

There are several minor speech communities within the Dravidian family. Notable among them are: Yurukala, Yerava, Tulu, Coorgi, Gondi, Malto and Kurukh-Oraon. The Kurukh-Oraon and Malto are confined to Bihar. They belong to the northern branch of the Dravidian family. Gondi, which is classified as a language of the central Dravidian branch, is the traditional dialect of the Gonds. However, recent census data show a steep decline in the numerical strength of the Gondi speakers. At the 1991 census, the Gondi-speaking population numbered just 2.12 million. This is an indication of the loss of language due to assimilation into the dominant regional languages. More than 90 per cent of the Gondi (mother tongue) speakers live in Madhya Pradesh (70 per cent) and Maharashtra (21 per cent). The remaining population is found in Andhra Pradesh and Orissa. In other states, their number is too small.

Indo-European Languages : Both in terms of numerical strength and the territorial extent the Indo-European languages surpass all other language families in India. The speech territory extends from Rajasthan in the west to Assam in the east and from Jammu and Kashmir in the north to Goa in the south. In fact, the domain of the Indo-European family extends beyond the borders of Rajasthan on the west and continues over adjoining Pakistan. Notable among the languages of this family are Hindi, Bengali, Marathi, Urdu, Gujarati, Oriya, Punjabi and Assamese. Keeping in view their importance as many as 13 of the Indo-European languages have been included in the Eighth Schedule of the Indian Constitution. While Hindi and Urdu are spoken across many states, including southern states, other languages, such as Bengali, Marathi, Gujarati, Oriya, Punjabi and Assamese, are specific to their own states. For example, while Bengali is specific to West Bengal, Marathi and Gujarati have their domain in Maharashtra and Gujarat respectively. The dominance of Hindi is evident from the fact that there were 337 million speakers who claimed it as their mother tongue in the 1991 census. About 81 per cent of the total population of Bihar, 91 per cent of Haryana and 89 per cent of Himachal Pradesh claimed affinity to Hindi.

The respective percentages for Rajasthan and Madhya Pradesh were 89 and 90. The story of the Hindi-speaking population is rather complicated. There are no less than 50 dialects which are grouped under Hindi. The speakers of these dialects declared them as their mother tongue in the same way as millions of others accepted Hindi as their mother tongue without making any reference to the dialect used by them. These dialects are actually regional variants of a spoken language of which the standardized form written in the Devnagri script is the official Hindi. Three-fifths of all Hindi speakers are concentrated in the two northern states of Uttar Pradesh and Bihar. Of the remaining, about 17 per cent live in Madhya Pradesh and 12 per cent in Rajasthan. The rest of the Hindi-speaking population is found in Haryana, Delhi, Himachal Pradesh, Maharashtra and West Bengal. In terms of numerical strength Bengali comes next to Hindi. While its speakers are heavily concentrated in West Bengal, a sizeable proportion of Bengali speakers is also found in the neighbouring states of Assam, Bihar and Tripura. Marathi is next to Bengali. About 93 per cent of its speakers live in Maharashtra alone. However, Marathi is also spoken by a section of population in Karnataka as well as Madhya Pradesh. Urdu holds the fourth rank among the Indo-Aryan languages. Its core region overlaps with that of the Hindi.

Its territorial extent is as vast as that of Hindi, although there is no state, barring Jammu and Kashmir, where it enjoys the status of the official language. About one-half of all Urdu speakers are concentrated in the states of Uttar Pradesh and

Bihar. But the strength of Urdu speakers is also sizeable in Maharashtra, Andhra Pradesh and Karnataka. A section of population in Tamil Nadu also claims Urdu as mother tongue. Gujarati has its domain in Gujarat, although other dialects, such as Gujrao/Gujrau and Saurashtri, are also claimed as mother tongues by a sizeable section of the population of Gujarat. Migration and settlement in neighbouring Maharashtra have resulted in the emergence of Gujarati speakers in that state. About 93 per cent of Oriya speakers live in Orissa alone, although the language is also spoken in the neighbouring parts of Madhya Pradesh and Bihar as well as Andhra Pradesh, West Bengal and Assam.

While the domain of Punjabi lies in Punjab, it is widely spoken over the entire northwestern region, particularly Haryana, Hi- machal Pradesh, Jammu and Kashmir and northern Rajasthan. Its territorial extent is wider as Punjabi is spoken in the neighbouring Punjab in Pakistan. Within India migration processes have taken Punjabi-speaking population to different parts of the country. In the east, almost 99 per cent of the Assamese speakers are confined to Assam. The domain of the Assamese is surrounded on the north, east, south-east and the south by Tibeto-Burman and Austric languages. There is a slight spill-over of the Assamese population into the neighbouring states of Arunachal Pradesh and Meghalaya.

There are other Indo-European languages, such as Bhili/ Bhilodi, Konkani, Kashmiri, Nepali and Sindhi, which have their own specific pockets in northern and western India. An overwhelming majority of the Bhili/Bhilodi speakers is concentrated in Rajasthan and Madhya Pradesh. Together these states share more than 80 per cent of the total strength of its speakers. Of the remaining, 17 per cent live in Maharashtra and 1 per cent in Dadra and Nagar Haveli. The domain of the Konkani lies on the Konkan coast from Maharashtra to northern Kerala. More than 90 per cent of the Konkani speakers are confined to the states of Karnataka, Goa and Maharashtra. In fact, they mainly live in the littoral region in the neighbourhood of Goa. Some speakers of the Konkani have also dispersed to other neighbouring states, such as Kerala and Gujarat as well as the union territory of Dadra and Nagar Haveli. Since the 1991 census was not conducted in the state of Jammu and Kashmir, the home of the Kashmiri language, it is difficult to describe the patterns of geographic distribution of the Kashmiri speakers. Outside the state a population of 56,000 persons returned Kashmiri as their mother tongue.

The Nepali-speaking population is distributed in a number of states, mostly in the neighbourhood of Nepal. Of the 2.07 million Nepali speakers, more than 40 per cent are concentrated in West Bengal and another 21 per cent in Assam. A small proportion (12.35 per cent) of the Nepali speakers are also found in Sikkim. They are scattered all over the northeast although in small numbers. The domain of Sindhi lies in the Sind province of the neighbouring state of Pakistan. The present Sindhi-speaking population in India, however, consists of population displaced in the wake of Partition in 1947. Initially, the Sindhis came to the neighbouring regions of Rajasthan and Gujarat. Later, they dispersed to other western and northern states. At the 1991 census, about two-thirds of all Sindhi speakers in India were enumerated in Gujarat and Maharashtra. Of the remaining one-third, the two states of Rajasthan and Madhya Pradesh shared together 15 per cent each. A small proportion of Sindhi speakers is also found in Delhi and Uttar Pradesh.

Literary and non-literary Language

The language scene in tribal areas of India deserves a mention. During the 50 years since independence, the tribes have been exposed to diverse influences—economic, political and socio-cultural. The scene in north-eastern India is somewhat different. There, the tribes have been empowered to manage their own political affairs. In other regions of India a certain number of seats have been reserved for the tribes to ensure their representation in the state and the national legislative bodies. These measures have paved the way for their rehabilitation in the national polity. However, a majority of Indian tribes lives in the mid-Indian region where their participation in the political processes is nominal. Their traditional habitats lie divided between several states. They do not have much of a say in policy formulation.

This has left an imprint on their traditional culture, language and social structure. Language seems to have suffered the most. First, the developmental processes initiated since independence seem to have contributed towards the disintegration of tribal economies, and their communitarian way of life. The tribes have lost their hold on the forest resources and have been forced to depend on the market forces. In fact, the free market economy has encircled completely the petty tribal commodity trade. Secondly, expansion of primary education has brought tribal children face to face to a new cultural situation. In the course of schooling they have been exposed to the regional languages of the states of their habitation. This has paved the way for their becoming bilingual. The ultimate effect is on their traditional dialects which are on the way to decline and eventual death.

It has been noted that the Indian tribes display a very high degree of diversity in their language affinity. Despite the relative isolation of the tribal communities there have been contact areas in which give-and-take between the tribal and non-tribal languages has continued throughout history. The geographic patterning of tribal languages suggests that along the zone of contact between them and the non-tribes progressive interaction has resulted in the fusion of linguistic elements on either side. This is evident from the fact that while the tribes communicate mainly in the Nishada, Kirata and Dravidian languages, they have also adopted several speech-forms of the Indo-Aryan family. The incidence of bilingualism and multilingualism among the tribes has increased phenomenally.

One may develop an understanding of the linguistic plurality observed in tribal areas by selecting the case of Austric-speaking tribes. In India, the Austric-speaking tribes are grouped into Mon-Khmer and Munda branches of languages. We have already seen that the Munda-speaking zone extends over a vast area from the Aravallis in the west to the Raj Mahal hills in the east.

Lingual Transition

A striking feature of the language scene in tribal areas is the growing shift in language affinity of the tribal communities. This fluid situation in which the tribes are losing their linguistic identity and are being identified with languages spoken by other tribes or the dominant regional languages of the states in which they have been living is observed in many parts of India.

An evaluation of the census data reveals gaps between the numerical strength of the ethnic tribals, say the Mundas, Santhals and the Gonds and those sections of the Munda, Santal and Gond population who declared their own dialects as their mother tongue, say at the 1961 census.

This lack of conformity between ethnic identity and language affinity reveals the process of language shift in a significant way. There can be several plausible explanations for this phenomenon. It may be assumed that by 1961 a major shift in the linguistic/dialectal affinity of the Indian tribes had already taken place in certain regions of the country. The 1961 census may, however, be taken as a benchmark.

It may be assumed that as tribal/non-tribal interaction was growing, a section of the tribal population shifted to other dialects/ languages with which it has no traditional affinity. This shift to the dominant languages of the regions of their habitation indicated a process of language shift and assimilation into the regional languages.

The language shift was, however, not necessarily from a tribal to a non-tribal dialect. In fact, several tribal groups shifted over to the other tribal dialects as contacts between them were growing fast. As a result they lost their own traditional dialects. The 1961 census data on linguistic affinity of the tribal communities as revealed in their declaration of a particular language as their mother tongue makes it possible to analyse the following dimensions of language shift among the tribes:

(a) Tribes speaking a dialect with which they are traditionally identified. For example, Santhals declaring the Santali as their mother tongue or Gonds declaring Gondi as their mother tongue. This reveals that the tribes in certain regions of the country display a continuity in their language affinity. We may call it a case of language retention.

(b) Tribes declaring a regional language as their mother tongue. For example, Santhals declaring Bengali as their mother tongue in West Bengal, or Gonds declaring Oriya as their mother tongue in Orissa.

This shows the ongoing process of language shift indicating that the tribal communities are getting assimilated into the dominant regional languages. Tribal regions where the process of regional development has brought tribes face to face to non-tribes have witnessed this phenomenon more significantly,

(c) Tribes on the periphery of their traditional areas have a tendency to declare as their mother tongue a language which is spoken by a dominant tribal group or the official language of a neighbouring state. This process indicates that the tribes are getting exposed to other tribal/regional languages. As a result they are getting assimilated into these languages (Table).

Language Shift

Tribes as % to District Population	*Radical*	*Important*	*Significant*	*Marginal*	*Insignificant*
80-100	The Dangs		Kohima Mokokchung	Jhabua	United Khasi and Jaintia Hills, Tuensang
50-80	Mandla	Lahul and Spiti, Dhar,	Mayurbhanj, Sarguja, Sundergarh	Ranchi, Bastar Koraput,	Mikir Hills, Kinnaur, Dungarpur, Banswara
20-50	Panch-Mahals, Vadodara, Bharauch, Surat, Shahdol, Sidhi, Nasik, Thane, Sambalpur, Sirohi	Khargone, Chindwara, Seoni, Kalahandi, Udaipur	Raigarh, Keonjhar	Batul, Dhulia, Baudh-Khondmals	Santal Parganas, Singhbhum, Chamba, Sawai Madhopur, Jalpaiguri
10-20	Visakhapatnam, Sabarkantha, Panna, Satna, Rewa, Raisen, Bilaspur, Durg, Dhenkanal, Bolangir, Chittorgarh	Khamman, Kamrup, Darrang, Palamau, Raipur, Yeotmal	Adilabad, Lakhimpur, Hazaribagh, Ratlam, Chandrapur, Purulia	Goalpar, Dhanbad, Bundi, Jhalawar, Darjeeling, West Dinajpur, Bankura	Jaipur, Kota
5-10	Nellore, Nowgong, Banaskantha, Shivpuri, Guna, Hoshangabad, Kolaba, Ahmad Nagar	Srikakulam, Sibsagar, Dewas, Jalgaon, Balasore	Khandwa, Coorg		Amravati, Ganjam, Alwar

Note: The language shift is described here as *radical* if 80-100 per cent of the tribal population have declared a regional language as their mother tongue. The shift is *important* if the proportion of tribes declaring a regional language as their mother tongue is 50 to 80 per cent. It is *significant* if the proportion is between 20 and 50 per cent. The shift is *marginal* if the tribes claiming a regional language as their mother tongue have a proportionate share between 5 and 20 per cent. The shift is *insignificant* if their proportion is less than 5 per cent.

Lingual Conservation

While collection of data on mother tongues may replate with problems, the position as recorded by the 1961 census was that one-half of the tribal population of India retained their own dialects as mother tongues. This was an evidence of language retention. The situation, however, varied from tribe to tribe and from region to region. The geographic patterning of the tribes still claiming affinity to their own tribal dialects revealed three major formations:

(a) Areas of tribal fastness in which tribes were by and large living in a state of exclusivity. For example, in Mizoram, Manipur, Meghalaya, Tripura, Nagaland, Arunachal Pradesh, West Bengal, Rajasthan, and Himachal Pradesh, 70-100 per cent of tribes retained their traditional dialects as their mother tongue.

(b) Areas of tribal-non-tribal inter-mingling, where tribes were living in a state of varied degrees of exposure to non-tribal economic and cultural influences. One can cite the case of Assam, Orissa, Andhra Pradesh, Madhya Pradesh, Maharashtra and Mysore(now Karnataka) where 25-60 per cent of tribes retained their own language.

(c) Areas where tribes have been assimilated into dominant cultures of the regions of their habitation. In these areas less than 10 percent of the tribal population claimed affinity with their own traditional tribal dialect (Table).

A number of doctoral theses, written under the direction of this writer at the Centre for the Study of Regional Development of the Jawaharlal Nehru University, explored these questions at length. In two of these theses the question of language shift and retention as registered among the Austric-speaking tribes of the mid-Indian region was examined. These studies revealed that the Santhals and Korkus by and large preferred to retain their traditional dialect. On the other hand, there were other Austric-speaking tribes who displayed a tendency of shift to the regional languages. The language shift was the highest among the Savaras. The Kharias, Mundas and the Hos followed. Studies also revealed that the Bhils of Rajasthan, Gujarat and Madhya Pradesh preferred to claim regional language as their mother tongue.

A study of the household-level data generated through field-work from Wanera Para, Umedgarhi, Nai Abadi and Regania villages of Bagidora tehsil of Banswara district, as well as from Banswara town, revealed that the Bhils by and large maintained Bhili as their mother tongue. On the other hand, the Korkus of Punasa, Richhi and Udaipur villages of Khandwa district tended to switch over to the regional languages. In fact, they declared Nimadi as their mother tongue. However, no generalizations can be made about Korkus on this basis as in other villages they continue to retain their own dialect and declare it as their mother tongue at the successive censuses.

Geographic Patterns in Language Retention

Proportion	*Areas*	*Tribal Communities*
90-100	United Mikir and Kachar Hills	Garo, Khasi, Jaintia, Kachari, Mizo, Mikir, Hmar
	United Khasi and Jaintia Hills	Garo, Khasi, Jaintia
	Tuensang	Naga
	Alwar, Bharatpur, Sawai Madhopur, Jaipur, Tonk	Mina
	Kota, Dungarpur, Banswara	Bhil
	Amravati, Batul	Gond, Pardhan, Kamar, Halba, Sahariya
	Santal Parganas, Singhbhum	Santal, Oraon, Munda
	West Dinajpur, Malda, Birbhum	Mahli, Kora, Mal-Paharia, Lohara, Kharwar

Contd...

Proportion	*Areas*	*Tribal Communities*
	Ganjam	Savara, Khond, Lodha
	Nilgiris Kammayakan	Kurumba, Palliyan, Pariyan, Kadar,
70-90	Hazaribagh, Ranchi, Dhanbad	Santal, Oraon, Munda, Lohara, Kharwar, Bedia
	Burdwan, Bankura, Midnapore	Santal, Oraon, Munda, Kora, Bhumij, Mahli
	Jhabua, Khandwa, Bastar	Bhil, Gond, Halba, Bhattra, Sahariya
	Goalpara, Cachar	Boro, Kachari, Miri, Rabha, Barman
	Baudh-Khondmals,	Konda-Dhora, Gond, Gadaba, Bhuiy, Kolha,
	Mayurbhanj	Bhumij, Santal, Ho
	Bundi, Jhalawar	Mina, Bhil, Varli, Padhi
	Dhulia, Chandarpur	Bhil, Naikdu, Dhanka, Gamit
50-70	Mikir Hills	Mikir, Dimasa(Kachari), Kuki, Hmar, Hajong
	Ratlam	Bhil
	Kohima, Mokokchung	Naga, Dimasa, Kuki
	Keonjhar, Koraput	Gond, Koya, Savara, Konda-Dhora Gadaba, Parja, Khond, Jatapu, Lodha, Gond, Munda, Bhuiya, Kolha, Bhumij, Santal, Ho, Kol, Saunti, Kissan, Mundari, Bathudi, Jwang
	Purulia	Santal, Oraon, Munda, Bhumij, Mahli, Kora, Malpaharia, Lohara, Korwa, Savara

Another finding of this research is that tribes living in rural areas have a greater affinity with their traditional dialects as compared to the tribes in urban areas. In regions of tribal concentration, for example, tribes enjoyed a certain degree of isolation which helped them retain their language and culture. Their stay in cities and towns, on the other hand, diluted their cultural identity and their language was the first casualty.

Investigations at the household level confirmed the language shift among the Mundas of Ranchi town and the Korkus of Khandwa tehsil, East Nimar district. On the contrary, the Bhils of Banswara district, the Santhals of Santal Parganas, the Korkus of Khalwa tehsil (East Nimar district) and the Mundas of the rural parts of Ranchi district have continued to retain their language.

The study noted a strong correlation between language shift and a number of associated factors, such as urbanization, proportion of Hinduized tribes, and the proportion of non-primary workers. Literacy also revealed a high positive correlation with language shift. In fact, the school-going tribal children were receiving education through the medium of regional languages. Obviously, schooling became a powerful instrument of bilingualism and/or language shift. In terms of exposure of the mid-Indian tribes to regional languages, Bhils stand first, followed by Mundas, Korkus and Santhals. In any case, loss of language is symptomatic of the loss of cultural identity. While the incidence of bilingualism among the Indian tribes is very high, it does not mean that they have always retained their traditional languages.

This is understandable in view of the fact that most of the tribes are getting exposed to external influences, particularly at the schools and the market places. Interaction at these places is possible through a common language, generally a regional language or pidgin, such as the Sadan/Sadari in Ranchi, which is adopted by the tribes for day-to-day communication. Back at their homes, their own dialect reigns supreme. This may not be true for the displaced tribals whose number is increasing day by day.

7

Geography and Caste System

Indian Caste Formation

Caste is a basic attribute of the Indian social structure. For centuries it has served as a major reference point in social interaction and continues to do so in some form or the other even today. Social hierarchy is based on caste and it is this philosophical vision that determines the behaviour of millions of Indian people in whichever walk of life they are. This is not to say that the caste attitudes are not undergoing changes as a result of education, economic development and social and political reform movements. The impact of these movements has been particularly significant since independence. Caste rigidities have weakened with time in the course of economic development, urbanization, weakening of feudal values, and more importantly, in the process of building a secular, democratic polity. Land reforms, coupled with other legislative measures to rehabilitate the socially deprived and the underprivileged sections of population, have helped transform the traditional view of caste.

Urbanization and industrialization in many regions of the country have resulted in occupational diversification thus changing the traditional inter-caste relations drastically. Migration has created a new climate for change with people adopting new occupations and leaving behind their home villages and traditional occupations which were caste specific. In cities, like Delhi or Kolkata, the caste antecedents of an individual are not immediately known and cannot be easily ascertained. This does not mean that with all these changes caste as an institution has died out or its hold on the society has been loosened. On the contrary, the introduction of democratic institutions and people's representation in elected bodies from the *gram panchayats, zila parishads,* municipal committees, state assemblies to the national parliament, have given a new meaning to the caste identity.

The geographic patterning of the caste groups is such that in some constituencies some castes are preponderant. This obviously gives them an upper hand in electoral politics as evident in the fielding of candidates of the same caste by all political parties. This results in the mobilization of votes along caste lines. Land reforms introduced after independence brought about a new equation between the castes. Those castes which had remained traditional owners of land by and large lost their monopoly on land. The land rights were transferred to their tenants who were mostly middle ranking castes. This naturally resulted in the dilution of the supremacy of the traditional landed elites.

They had to face new challenges from sections of population who had been subservient to their interests so far and had been living under their perpetual hegemony. The state also adopted a policy of scheduling certain socially deprived castes. These castes of the scheduled category enjoyed certain privileges. For example, state and parliamentary constituencies were earmarked and reserved for them, they had a quota in government jobs, and scholarships were instituted for them to enable them to join schools and higher institutions of learning. These opportunities were new to them and were intended to help them overcome their traditional drawbacks. However, this also disturbed the social equilibrium which had sustained the Indian society through ages. Higher castes and the middle ranking castes were generally wary of these reforms and openly contested these policy initiatives. Thus caste in independent India became a volatile issue. The area of inter-caste struggles widened and the violence that followed was a new phenomenon unprecedented in Indian history.

It may be necessary to examine these issues at some length and map out the socio-geographical outlines of the caste phenomena in contemporary Indian society. The changing role of caste in politics, economy, education, social networking and bureaucracy may be probed in order to comprehend the existing social reality.

Traditionally, the Indian society has been governed by the institution of caste. This institutional frame has operated through history as a constraint on the freedom of an individual. The position of an individual in society and his/her sphere of social interaction and code of conduct were determined by his/her caste, that is to say, the accident of birth in a family. Caste, therefore, operated as an instrument of social ordering. Each individual was supposed to lead a life in accordance with the socially accepted norms imposed by this rigid and unchanging social order defined by the caste. The concept of caste should be evaluated in this historical backdrop.

The Genesis

It is generally believed that as a phenomenon caste is unique and a specificity of the Indian society. Tracing its origins may, therefore, be necessary to understand the social mechanics that led to the emergence of a hierarchized social order based on caste. The origins of the caste system are, of course, shrouded in mystery, although there is a general agreement that the beginnings were made in the Vedic Age. It all began as a part of the process of peopling of the Sapta-Sindhva region by the Indo-Aryans after the decline of the Indus valley civilization. The racial composition of population seems to be important. While the Palaeo-Mediterraneans and the Proto-Austroloids were already there, a new ethnic element that was introduced in the population was that of the Nordics (Indo-Aryans). As the mode of agriculture took its roots and expanded in space it resulted in a new organization of social forces of production.

The expansion of agricultural mode implied assimilation of hunting and food gathering as well as pastoral communities into an agricultural society. This process of transformation was attended by a simple division of labour among the different constituent groups who came together to live in the pioneer villages. Understandably, the colour of the skin became the primary basis of this social division. The Indo-Aryans asserted their superiority by creating a four-fold division of the society which came to be known as *Chatur Varnya* and placing themselves in the upper strata of the Brahman and Kshatriya ranks. The Indus valley people and the tribes who lived on the periphery were assigned a lower status of Vaisyas and Sudras, or even treated as out castes.

The Indo-Aryans emerged as the custodians of knowledge (Brahmans) and defenders of territory (Kshatriyas). On the other hand, the Indus valley people were assigned mundane jobs such as cultivating the fields or exchanging and manufacturing artisan goods. There were other unclean jobs, such as treating the dead bodies or removing the skins of the dead animals. These jobs were handled by the lower strata of the society, sometimes described as the 'exterior' castes.

It may be interesting to note that while a similar division

of labour emerged in all other agrarian societies in the world, an elaborate caste system emerged only in India. Why India is an exception to this rule? What led to the institutionalization of this division of labour which became so rigid that it could not be altered or challenged? Each caste group became associated with a specific department of work, so much so that a particular vocation became a hereditary trait of the family, immutable in time. It is generally known that initially a four-fold *varna* division emerged in the Vedic Age.

These four *varnas* were Brahman, Kshatriya, Vaisya and Sudra. These groups constituted a hierarchical social division with Brahmans at the top and Sudras at the bottom. Within each group, particularly among the two upper segments of the *varna* system, individual families started getting differentiated on the basis of clan or *gotra*. These divisions became institutionalized as marriages were permissible within the same *varna*. On the other hand, it was not customary to marry within the same *gotra*. This gradually led to the emergence of a social practice of *varna* endogamy and *gotra* exogamy.

Varnas were also differentiable from one another on the basis of rituals performed on occasions such as birth, marriage and death. The details of these rituals were prescribed in the scriptures (*Brahamanas* and *Smritis*).

For all practical purposes *varna* acquired all the characteristic features of a class division. In the social system of *varna* the Brahman and the Kshatriya emerged as the upper classes. The Brahmanical class monopolized knowledge and imparted it to the children of the upper classes on a selective basis. The Brahmanical class also legitimized the role of Kshatriya in enforcing the law, thus enabling them to rise as a ruling class. The Kshatriyas naturally assumed their role as defenders of territory. The Brahmanical interpretation was that the Kshatriya was a legitimate *raja*, and more importantly, a divine incarnation.

Later on, as economy diversified and a section of population concentrated in large settlements, occupations also diversified. Eventually, the *varna* divisions ramified in accordance with this process of economic diversification into the *jati* divisions. The *jatis*, however, were hereditarily associated with certain occupations which defined their ritual status and the rank in society. The *jati pratha* was broadly in conformity with the *varna* system.

This means that each *jati* corresponded to a particular *varna*. The ranking of the *jatis* was relative to one another and in accordance with the *varna* order. The *varna-jati* system was also specific to the cultural regions and inter-regional comparsons were futile. It may be concluded that the *varna* system acted as an all-India model representing the Indian unity. *The jati* system, on the other hand, was a regional specificity. The question of relations of one caste with the other caste and the relations with the members of the same caste can perhaps be understood in the context of regions.

The *jati* system, thus, represented the diversity of the regional ethos within the over-arching all-India unity. It may be noted that the process of cultural differentiation expressed itself over a domain of language, implying that each cultural region was in reality a language/ dialect region. Each region was differentiable from neighbouring regions on the basis of spoken language. A particular *jati* strictly functioned within a language area. The *jatis* in Bengal are not comparable with the *jatis* in Karnataka, or for that matter, in Maharashtra. This is not to say that migrations from one region to the other have not been common.

In fact, migrations were a common phenomenon. They did result in a mechanism of preservation of cultural traits. Each emigrant group in a city found it more convenient to function as a brotherhood. This promoted intra-group interaction. The immigrant groups, divorced as they were from their traditional homelands, evolved a system which kept them as a closely-knit community, rooted in their cultural tradition and yet living in an alien region. It was this mechanism that made inter-regional (cross-cultural) marriages a rare phenomenon. With the passage of time sub-castes which were splinter groups from the major castes came into being. While initially, they were sub-divisions of a caste, gradually each sub-caste started functioning as a caste in itself. Within the sub-castes a system emerged which imposed restrictions on marriage, commensal relations and communal behaviour.

The sociological literature on caste is voluminous. Studies on caste cover a vast area in Indological, social-anthropological and sociological research. Obviously, the approaches vastly differ. It may be pertinent to refer here to the differing viewpoints on the origins of *varna-jati* system. Briefly, three major viewpoints can be highlighted here. First, caste is viewed as a product of the division of labour. Secondly, it is viewed as a cultural phenomenon and operates as an institution. Thirdly, caste is viewed as social stratification leading to the institutionalization of social inequality. It is this vision of social inequality ingrained in the Indian mind that confronts with the principles of democratic polity introduced since independence.

For all practical purposes membership of a caste is hereditary and cannot be acquired. Membership of a certain caste also determines the marriage field as all caste groups are endogamous. It also determines the pattern of social interaction and commensal relations (whether one can accept or can not accept cooked or uncooked food or water from a number of a certain caste).

Caste-based Set-up

The pioneer villages that emerged in the Sapta-Sindhva region initially and the Ganga and the peninsular river valleys later evolved a social division based on the caste system. In fact, the members of the different caste groups came together in a certain proportion depending on the area of the cultivated land, productivity of soil and a certain mix of crops. As caste groups converged on land reclaimed from the forest the territory was dotted with pioneer villages of different lineage groups. Understandably, cultivators constituted the nucleus of the village community.

The cultivating castes might have been self-sufficient in the beginning. But, as the process of agricultural intensification progressed, they needed other people who could help them in agricultural operations or provide services, such as those of a potter, a carpenter, a barber, a weaver, a shoe-maker and the like. These groups who came together in a village setting specialized in specific rural arts. Their number was, however, determined by the demand posed by the cultivating caste. In a typical village setting, there were weavers, potters, carpenters, blacksmiths, shoe-makers, oilmen as well as those who treated the dead bodies at the cremation grounds or treated the skins of the dead animals for making leather.

They formed a ring around the nucleus of the cultivating castes. As is evident, these services were not, and could not be, treated as of equal rank. These groups were arranged along the vertical scale with an implicit hierarchy. Each one of them constituted a *jati* based on a specific occupation with which it had a traditional association. Those who handled unclean and polluted jobs were drawn from the Sudra ranks and were mostly treated as untouchable. It may be true to say that this model was applicable to the pioneer villages that emerged in the first millennium B.C. in the Ganga valley and a little later in the valleys of Krishna, Godavari and Kaveri in South India.

The village mirrors the intricacies of the caste-based Indian society. The ritual hierarchy of castes expresses in space in the residential patterning of different castes within a rural settlement in whatever region of India it is located. In fact, the social morphology of rural villages in India is a spatial surrogate of the ritual hierarchical social order based on caste. What happened was that the most favourable sites within the village settlement *(abadi)* were occupied by the cultivating caste, the *dominant* caste.

The Brahman houses were located in their neighbourhood. Separated by lanes were the quarters in which the middle ranking castes lived. The outer periphery of this agglomerated settlement unit was occupied by other service castes that generally performed clean services. There was yet another periphery in which the unclean castes, or the untouchables, holding the lowest rank in caste hierarchy, were living. Depending on relief as in Rajasthan, or if the settlements were perched on mounds as in parts of Uttar Pradesh, the central promontory was occupied by the main cultivating caste, say Rajputs, Jats or Ahirs.

At the lower level were other middle ranking castes and the low castes were situated on the outer periphery which was generally a low-lying area. Thus, an altitudinal zonation based on caste hierarchy characterized the rural settlement. This model was universal and represented a type in itself. If Muslims, *e.g.*, Sheikhs, Julahas or Fakirs, were a component of rural population, they also followed a zonal distribution pattern similar to that of the Hindu castes.

O.H.K. Spate noted that an Indian village did not follow a definite layout plan. However, "within the seemingly chaotic agglomeration there is, as a rule, a strong internal differential, that of the separate quarters for various castes" . His study of Aminbhavi, a village situated to the north-northeast of Dharwar in Karnataka, although randomly chosen, was a typical example of the way caste factor operated in the residential patterning of different communities in a village. Based on a pioneering survey undertaken by CD. Deshpande and his students and with subsequent inputs from L.S. Bhat, Spate's study can serve as a model.

The village population consisted of Lingayats, an agricultural caste of Karnataka (64.54 per cent), Muslims (13.40 per cent), Jains (6.09 per cent) and Brahmans (1.83 per cent). The members of each caste occupied a solid block of contiguous houses in a lane named after the caste. The village population also included Talwars (7.31 per cent), Harijans (4.87 per cent), Wadars, Shikalgars and washermen. On the periphery of the village lived the low castes. The occupational structure as in other parts of India is caste-based. For example, Lingayats are mostly tenant farmers whereas Talwars and Harijans are generally agricultural labourers. Members of other castes are carpenters, smiths, cobblers, washermen, barbers, etc. Despite the changes in Aminbhavi, the residential patterning continues to be the same.

While caste-based residential segregation is universal, the village in south (India) presents an extreme form of spatial segregation.

"The general emphasis on caste in the south (India) takes social fragmentation, allied with spatial separation, to the extreme, segregating the untouchable in outlying *cheris* or sub villages, sometimes located several hundred yards from the main villages of which they are service-components. This is indeed a climax of geographical differentiation; apartheid" (Box).

> *A Typical Chert :* A typical *cheri* may consist of two rows of huts with a narrow central 'street'; in the middle this widens to make room for a tiny temple. The huts have thick mud walls, roofed with palmyra thatch, and low mud porches scrupulously swept. To enter one must bend double; the only light comes from the door and from under the caves,

> and furniture consists of a few pots and pans, a couple of wooden chests, and the essential paddy bin, ... raised from the ground to escape the rats and built up of hoops of mud. Poor as they are, these dwellings are yet homes, and obviously loved as such: their cleanliness, the surrounding mangoes, coconut and palmyra palms, redeem them from utter squalor. The nadir is reached in the *bustees* of Calcutta and the revolting camps of casual tribal labour found on the outskirts of the larger towns: shelters (they cannot be called even huts) of matting, of rags, of petrol tins beaten flat, on the waste spaces open to the sun and racking with filth.
>
> —O.H.K. Spate, *India and Pakistan: 1957*, 203-04

Andre Beteille's study of Sripuram village of Tanjore district in Tamil Nadu shows that:

"people who are close to each other in the social system tend to live side by side; people whose social positions are widely different live apart. The *agraharam*, for instance, is not only a cluster of habitations but also the centre of social life for the Brahmins... to the Brahmins, the *agraharam*, in more ways than one, is the village" .

In the same way the *pallacheri* is not just another quarter of the village. It is a place in which the entry of the Brahmans is forbidden by social custom. They just cannot enter the *pallacheri* since the concept of pollution and purity applies to places also. It is not just interpersonal.

Castes in Urban Neighbourhood

The residential housing in urban neighbourhoods is a continuation of the rural model—separate quarters for different castes. Take, for example, the towns of western Uttar Pradesh, such as Mathura, Aligarh, Bulandshahr, Meerut, Moradabad, Bareilly, Badaun and Firozabad. There are separate *mohallas* for different Hindu castes. And this pattern is valid for all regions of India. If there are Muslims in these towns they live in separate *mohallas, katras* or *bustees.* In the towns of British India there were mixed localities as well in which Hindu castes lived in the midst of Muslims or vice versa. However, segregation has increased with time and in many North Indian towns different communities live in separate blocks.

In a traditional set-up, there are separate residential quarters for Brahmans (*e.g., mohalla* Chauhan in Firozabad), Vaisyas, blacksmiths, weavers, washermen, tailors and other artisan castes. The low castes such as the sweepers and the Kanjars generally live on the periphery of the town. There is an association between the *mohalla* name and the castes that live in them. This is true for Muslim *mohallas* as well. There are separate quarters for separate castes (*e.g., katra* Sonaran, *katra* Pathanan, *mohalla* Shishgaran in Firozabad, *mohalla* Churiwallan in old Delhi).

With development these towns have grown into large cities with new outgrowths mushrooming on the periphery but the caste-based segregation is still intact and the community-based segregation has further widened. It may be noted that in the new Muslim residential colonies caste-based segregation is generally missing. However, propinquity between the houses of a Pathan and a Julaha in a new colony does not necessarily mean that there is an enhanced social interaction between the two families or a drastic behavioural change in the social attitude towards each other.

The neighbourhoods in the Mughal town of Shahjahanabad presented yet another example of caste-based residential segregation. In the beginning, while the royalty lived in the fort located on the banks of the Jumna (Yamuna), there were the *mohallas* in the neighbourhood of the fort for the nobility in the area between the fort and the Jama Masjid. This building complex was razed to the ground after the uprising of 1857.

Around the Jama Masjid were arranged the *mohallas* of the service castes, as well as those of ethnic groups such as *mohalla* Punjabian, Kashmiri *mohalla, kucha* Pandit, Bara Hindurao and so on. Shahjahanabad continues to have the same pattern of neighbourhoods even today. There are *mohallas* with occupational distinction, such as Ballimaran, Suiwalan, Churigaran, *kucha* Chelan, Gali Rajan, etc. After the Partition of the country in 1947, Shahjahanabad changed a lot but the traditional *mohalla* structure is still intact.

Studies of the Himalayan towns such as Chamba, Kangra and Kullu reveal a close relationship between *mohalla* names and the caste/ethnic groups associated with them. In fact, the *mohalla* formation was essentially based on the principle of caste/class segregation. The capital towns of the princely states also indicated a clear impact of the feudal social order. The royalty and nobility chose for themselves the most advantageous sites, close to the fortress or the palace, followed by *mohallas* in which the elite groups (courtiers, officers, religious leaders) lived; next were the *mohallas* of artisan and trading castes (metal and stone workers, weavers, carpenters, masons, perfume workers).

Adjacent to them were the *mohallas* of service castes, both ritually pure (water carriers, domestic servants, cattle-keepers, shepherds) as well as impure (sweepers, cobblers, butchers). In a hilly town, an added dimension is that the caste hierarchy expresses itself in the altitudinal zonation of *mohallas: the higher the position in caste rank, the higher the location in the town, other things remaining the same. Mohallas* of Chamba such as Bansi Gopal, Nand and Drobhi have traditional habitations of the Brahman, Mahajan and Rajput families. Julahas (weavers) are concentrated in a *mohalla* called Julakri, smiths in Sapdi and butchers in Kasakra. The general pattern is that the *mohallas* of high status castes are surrounded by castes which have a lower rank in ritual hierarchy. On the periphery are the *mohallas* of the low castes.

Caste Influence on Behavioural Attitudes : As already noted, rural settlements in India have largely been influenced in their residential patterning by the caste and the clan/lineage structures. In fact, the caste influences in subtle ways both the behavioural attitudes of the people as well as the social values that determine them. As is generally known, the Indian village is a complex entity. While its evolutionary history has left its imprints on the layout of habitations, even the direction in which the houses of certain caste groups are situated and the daily movement patterns of the people from their homes to the fields and the market place are determined by caste factors. Moreover, caste does not function alone. It creates a network of relationships based on economic/functional inter-dependencies, ritual and religious customs and the feudal social order.

Every settlement in India is in a way a mosaic where people live in their own exclusive social worlds despite the fact that they meet members of other castes in different situations—secular as well as religious. The traditional way of life and the spatial organization of habitations continues without much of a change. However, democratic institutions to which the rural population has been exposed since independence and the impact of planned development including creation of social infrastructure within the villages (*e.g.*, schools, dispensaries, seed depots, banks, etc.) have added new elements to the traditional settlement morphology. These developments notwithstanding, the traditional order based on caste structure is still intact.

Studies conducted by Indian and foreign scholars in different parts of India confirm all-embracing role of caste and *jati* identities in the settlement morphology. The roots of this relationship between caste and settlement morphology have been traced in the early colonization of the Ganga valley and the emergence of territorial units *(janapadas)* in ancient India.

Such a complex social system as the caste/*jati* structure and its diverse regional facets was bound to have its expression in the social organization of rural villages and the territorial organization based on clan/lineage connections that arose out of it. Members of the same caste/*jati* tended to live together in the rural space out of necessity and the character of the locality came to be defined by the day-to-day cooperation between them emanating from the principles of common descent and economic bondage. Thus, the common descent and the common residence were associated in space. The situation continues to be the same even today.

Naturally, the kinship ties have expanded in space. Initially, they were localized, that is to say confined to the same village, but eventually they have gone beyond the village and have resulted in the emergence of caste or clan territories, loosely defined but firmly asserted by the lineage groups. The kinship ties which also operate through marriage networks further reinforce the kinship feeling. Thus, a spatial structure of great social and functional relevance based on lineage is created. K.N. Singh has cited the case of Rajputs, Jats, Bhumihars, Ahirs, Kurmis and Kansar Pathans living in the North Indian Plain who are known to have established a chain of clan territories through their patrilineal kin bodies.

"Territorial organization and colonization normally proceed through the hegemony and control of dominant classes... which tend to earn through time great amount of political... power... from regional authorities without and social and territorial control within on a local level such as clan territories."

Over the clan territories lineage groups of dominant *jatis* wield great social authority. These clans enjoy a kind of territorial suzerainty, since they are the landed castes and their land ownership is a great factor in contributing to their social authority. This is not to deny the fact that the population composition within the supposed clan territory is heterogeneous. In fact, several *castes/jatis* live together and their population is often larger than that of the dominant caste. These territories generally encompass a certain number of villages.

In popular imagination these units are generally recognized as consisting of a cluster of villages. This cluster may consist of 96 villages *(chhanbe)*, 94 villages *(chauranbe)*, 60 villages *(sathi)*, 42 villages *(biayalisi)*, 24 villages *(chaubisi)*, or 20 villages *(bisi)*. Evidently, the extent of the territory as popularly perceived may be a function of the initial settlement of a dominant caste in a given region and its subsequent expansion in space. This historical fact acquires great significance in defining the functional authority of a caste over a territory. It has its own behavioural implications.

It may be concluded that the residential rural space in India has evolved through history within the over-arching influence of caste relations. The distinction between castes expresses itself in the segregation of population of different castes in different *dhanis, puras* or *purwas*. The dispersal of population of different castes in the *dhanis* of villages in Rajasthan is an example of caste-based residential patterning. In fact, the general principle is: the lower the rank in ritual hierarchy, the farther the location of that caste from the centre of the settlement. By the same logic, the out castes are supposed to live in separate quarters outside the main village. This is not to deny the fact that the different caste groups have been attracted by the dominant castes for the services they render and their social interaction and the daily give-and-take that goes on between them on a routine basis is based by and large on economic and religious factors.

System of Landownership

Historically, the factor of caste has operated as a major determinant of social status in many critical spheres of life.

One such critical area was the access to land, the main productive asset in an agrarian society, such as India. As noted above caste groups were surrogate classes and a division between those who owned land and those who did not could easily be perceived along caste lines. The upper castes invariably emerged as the landed elites. On the other hand, those in the lower rungs of the caste society were largely landless, working on the land estates of the ownership farmers.

It is understandable that in a vast country like India with diverse agro-climatic conditions there were regional differences in the types of land ownership and the tenurial relations. It is also understandable that land systems of British India, such as those described as *ryotwari, mahalwari* or *zamindari,* came in response to the diverse agro-ecological conditions, and the regionally diverse socio-political structures. In the course of time, on the *zamindari* lands, an Indian form of feudalism emerged in which the landed elites were the high castes while the labour supply came from the low castes who were treated no better than the serfs. Exceptions apart, there was, therefore, a broader correspondence between the ritual hierarchy of caste and the distribution of ownership rights on land. The landed elites emerged as *dominant* castes (Box).

> ***Dominant Castes :*** In many parts of rural India there exist castes which are locally numerically preponderant, own the bulk of the arable land, occupy a fairly high position in the ritual hierarchy, and wield power over the other castes. Examples of such castes are Jats, Ahirs and Rajputs in the north, Patidars in Gujarat, Marathas in Maharashtra, Kamma and Reddi in Andhra Pradesh, Lingayat and Okkaliga (Vakkaliga) in Mysore, Vellala and Gounder in Madras and Nayar in Kerala. Sometimes the dominance of a caste is decisive, all types of power being concentrated in it. At other times, however, the different elements of dominance may

> be distributed among several castes. In the first instance, the dominant caste wields great power over the others, while in the second there is likely to be a balance of power among the powerful castes.... the maintenance of law and order in rural areas even now depends to some extent upon the leaders of the dominant caste. They are the people who can punish errant individuals and ensure the maintenance of the caste codes. When a caste is politically and economically dominant, its religious position tends to fall in line with its secular position.
> — M.N. Srinivas, *India: Social Structure*, 1986: 19

They were drawn from among the ranks of the Hindu high castes, such as the Rajputs, Kayasthas and Brahmans as well as upper castes among the Muslims. The Mughal policy of awarding land grants to the high castes among the Hindus and to the upper sections of the Muslim society, *i.e.*, Ashraf, comprising Sheikhs, Sayyids and Afghans, was apparently aimed at creating a broad-based ruling class through which the surplus from land could be appropriated.

The British retained the system as such. In fact, they further strengthened it turning it into a rigorous exploitative mechanism to serve the best interests of the colonial state. The growing accumulation of land in the hands of the members of the high castes eventually reduced the artisans, Sudras and the other sections of the out castes to the level of paupers. In fact, all chances of their having access to land were blocked by the dominant castes. Their marginalization acquired a form which contributed immensely to the Indian poverty.

Mughals and Caste System

The Mughal chronicle, *Ain-i-Akbari,* mentions the names of the landowning castes *(qaum-e-zamindaran)* in each of the *mahals* of the Mughal provinces. On the basis of this information it is possible to infer the nature of relationship between caste and landownership. Using these data it is also possible to know the geographical distribution of the landed castes in different parts of the Mughal empire. This study is, however, restricted to 13 sarkars chosen randomly from the seven provinces of India under Akbar. The *sarkars* chosen include Agrah, Kalpi and Alwar (Subah Agrah), Awadh, Lucknow (Subah Awadh), Dehli, Saharanpur (Subah Dehli), Rechnau Doab (Subah Lahor), Bhakkar (Subah Multan), Bihar (Subah Bihar), and Ujjain, Garha and Chanderi (Subah Malwah).

Sarkar Agrah : Landowning castes included Gaur (a branch of Rajputs), Jat, Lodh, Chauhan Rajputs, Bhadoriya (a branch of Chauhan Rajputs), Thatthar Rajputs (Gujjars converted to Islam), Brahman, Sheikhzadah, Ahir, Panwar Rajputs, Sikarwar (a branch of the Badgujjar Rajputs), Sheikhzada Chishti, Sayyid and Badgujjar. However, despite this diversity of landed castes most of the land was owned by the different clans of Rajputs. Brahmans and Sheikhzadahs are mentioned in two *Parganas* only. Jats are mentioned in three *Parganas* and Sayyids in one only.

Sarkar Kalpi : Landowning castes include Rajputs, Kachhwaha Rajputs, Sengar Rajputs (a branch of the Agnibansi Rajputs) and Rajput Kumbi, Bais, Sheikhzada, Malikzada, Brahman, Afghan and Turkoman. The names of the Rajput clans are mentioned repeatedly thus indicating that most of the land was under the possession of Rajput clans who were the dominant caste. The names of Sheikhzada, Brahman, Afghan and Turkoman are mentioned in one *Pargana* each.

Sarkar Alwar : Castes mentioned include Khanzadah (a Meo sub-caste), Kachhwaha, Badgujjar and Chauhan Rajputs, other Meo sub-castes, Mina, Jat and Abbasi. However, land was mostly owned by Khanzadahs, Meos of other *gotras* and Rajputs. Mina, Sayyid, Baqqal, Abbasi and Jat are mentioned in *one Pargana* each.

Sarkar Dehli **:** Castes mentioned include Rajputs, Chauhan Rajputs, Badgujjar Rajputs, Chhokar, Ahir, Afghan, Gujjar, Ranghar, Jat, Brahman, Taga and Sayyid. While different Rajput clans are the main landed caste, there were many other caste-groups who were holding land in different parts of the *sarkar* of Dehli.

Sarkar Saharanpur **:** Landowning castes mentioned include Ranghar, Taga, Jat, Gujjar, Ahir, Brahman, Afghan, Turkoman and Sayyid. Apparently, there were several caste-groups who were equally dominant in this *sarkar*.

Sarkar Awadh **:** Brahman, different Rajput clans, such as Kumbi, Chauhan, Ghelot, Gaikwar, and Sombansi, Bais (newly converted to Islam) and Ansari are the castes listed in the *Ain*. Evidently, most of the land was held by the different Rajput clans including those who were recent converts to Islam. Ansaris are mentioned only in one Pargana.

Sarkar Lucknow **:** Castes mentioned include Rajputs and their different clans, such as Chauhan, Chandel, Gehlot and Bachgoti. Other castes mentioned include Brahman, Sayyid, Jat, Afghan, Sheikhzada and Ansari.

In fact, Rajputs were the main landowning caste followed by Brahmans, Sheikhzada and Bais. Afghan and Sayyid are mentioned only in one *Pargana* each.

Sarkar Bihar **:** Afghan, Brahman, Kayath (Kayastha), Sheikhzada and Rajput were the major landowning castes. All the five caste groups are listed under the different *Parganas* thus indicating a diversity of landed castes.

Sarkar Chanderi **:** Bagri, Baqqal, Khati, Dangi, Ahir, Brahman, Rajput Bundela, Kayasth, Musalman, Khichi, Dingi, Makhati, Baqqal and Khatri were the main landowning castes. However, the data do not indicate dominance of any particular caste. Thus, the region shows a fairly large diversity of dominant castes.

Sarkar Garha : Gonds (a well known tribe) are shown as the only landowning community.

Sarkar Ujjain : Among the main landowning castes are Ujjainia, Rathor, Soriah, Magwar (Meghwal), Deora (Rajput), Bais and Jadon (Yadu,Yadava). Different Rajput clans were the dominant caste over most of the area.

Sarkar Rechnau Doab : Bhatti, Khokhar, Chimah, Jat, Manhas, Balawariah, Bhutiyalah, Taral, Brahman, Sanawal, Langah, Gujjar and Hatiyalah were the landowning castes in the different Parganas of this *sarkar*. However, it appears that the Jats and their different clans were the most dominant over most of the Rechnau (Ravi-Chenab) Doab region.

Sarkar Bhakkar : Castes mentioned include Dharejah, Mehar, Rahar, Jahna, Bhatti, Sahejah, Harejah and Jaman. Insofar as the caste dominance is concerned the picture is quite diverse.

It is thus obvious that historically land relations in India conformed to the social reality of caste. The *Ain's* data further confirm this view.

Colonial Power and Caste System

The character of Indian feudalism registered a major change under the British. In fact, the British created multiple interests in land and the number of intermediaries which increased as the *zamindari* system took its roots, particularly in Bengal and Oudh, added a new dimension to the exploitative processes of the *ryot*.

The caste system which was a manifestation of the social hierarchical order conformed to this pattern of absentee landlordism admirably. Caste-groups of various ranks took positions of vantage in this pyramid of feudal exploitation (Table).

Multiple Interests in Land: Examples from British India

One Interest	*Two Interests*	*Three Interests*	*Four Interests*	
• The government	• The government	• The government	• The government	• The government
	• The ryot	• A landlord	• A landlord	• An overlord
		• Actual holders	• Tenure holders	• A landlord
			• The ryot	• Actual holder

Notes: An occupant with a definite title, not a tenant.

A *zamindar, talukadar* or a village body as in the *mahalwari* system.

May be a co-sharer.

A sub-proprietor.

Source : B.H. Baden-Powell, *Land Revenue and Tenure in British India.*

India at independence inherited this pattern of land relations. The Indian National Congress which struggled for India's independence always pleaded for drastic land reforms in order to ensure a better deal for the oppressed sections of the rural society. The *zamindari* abolition in the early fifties of the twentieth century was a part of the land reforms package introduced by the national government under Nehru.

Status Quo

One may conclude that the relationship between caste and land-ownership is not simple. The fact that among the landless are included most of the low castes, designated as scheduled castes, and most of the land is in the hands of the privileged sections of the caste society speaks volumes about the social forces that operate at the village level. This question of economic rehabilitation of the oppressed sections has received attention in many sociological and anthropological studies conducted at the village level (Box).

The Changing Village : Caste itself is relatively unimpaired. Everyone, including the Ganjam DISTILLERS and the Boad OUTCASTES, 'believes in caste'. They observe the rules of endogamy and inter-dining. Caste usages affect every aspect of their lives. The solidarity of caste-groups has not lessened. : But the local structure is undergoing a fundamental change. It is losing its politico-economic function. This function rested on the fact that productive resources were allotted according to caste, and control of these resources was organized by a localized structure of caste-group.

In one case the structure has adjusted itself successfully to the new economy. But in other cases it has failed. The new economy is widespread: caste-group organization is appropriate to a localized economy. The new economy demands social mobility: but untouchability is at once a cornerstone of caste and an insuperable obstacle to mobility. Localization and untouchability are the rocks against which beats the tide of the new economy.

The man of Bisipara today derived their right to a share in the resources of production not by membership of a caste but as citizens of India. In the village the hierarchy of caste-groups is no longer a complete reflection of economic realities, nor an adequate means of ordering political relations. Under pressure of economic change the political functions of caste are beginning to be taken over, as one might expect, by the ultimate political authority, the Government of India.

—F.G. Bailey, *Caste and the Economic Frontier*, 1972: 275

Few geographers have ventured in this area. Reference may, however, be made of the pioneering work of S.M. Mandavdhare who investigated the question of caste and land relations empirically by selecting a sample of fifty villages from the Marathwada region of Maharashtra. His data based on individual caste households show a broad conformity between the status of a caste-group and the proportionate share of land it holds. The data were collected for all the castes living in the sample villages. The sample included the major agricultural castes of Marathwada, such as Maratha, Wanjari, Mali and Lingayat as well as the scheduled castes, such as the Mangs, Mahars and Chambhars (local name for Chamars). Marathwada, as this study shows, presented a case in which the number of marginal and small farmers is quite high.

The proportion of large farmers among the scheduled castes is very low. The study further shows that land inequality among the castes is significantly high in villages lying in the rainfall scarcity zone. Similarly, in villages where the scheduled castes are politically organized, inter-caste inequality has considerably reduced. This shows that, other things remaining the same, political mobilization may help the oppressed sections in their economic rehabilitation.

Broadly speaking, an understanding of the social status of a caste group and its share in the village land cannot be properly developed without linking it to the *jajmani* system and its local variants which formed the basis of interdependent relationship between the landowning and the landless castes. The caste-based production relations in rural India thus reveal features of class-like formations where caste actually acts as a surrogate class. As actual tillers on land are very often the low castes who neither own land nor have a just share in the surplus generated by them, agrarian relations have remained ridden with conflicts.

Distribution in Geography

There is no data base available to enable one to analyze either the numerical strength or the geographic distribution pattern of different caste-groups living in India today. The successive censuses conducted in India since independence happily opted to discontinue the earlier practice of collecting and publishing data on caste. However, data were and are still collected during all the censuses on the scheduled category of population—castes as well as tribes. Understandably, these data are partial and not enough as they tell us only a part of the story. The data are grossly inadequate insofar as the question of the caste identity of the people of India is concerned in the context of today.

In a country where the affairs of the state have been influenced by caste and community identities, non-availability of geographically disaggregated quantitative data on caste may be a serious handicap to social science research. For example, a socio-geographic interpretation of caste is not possible in the absence of such crucial data. There is no gainsaying the fact that geographical distribution of caste-groups helps in understanding the character of ethno-lingual regions as well as the implicit meaning of the ongoing dynamics of social change in different parts of the country. Caste is a social reality. Its role in society and its politics cannot be minimized by suppressing data.

The levels of social development reached by different caste-groups are an important indicator of their progress and the place in society. The relations between different castes, as the contemporary evidence shows, are becoming increasingly competitive. This is largely due to the skewed distribution of wealth. Hidden behind these conflicts is the basic question of social justice. Moreover, caste identity is an important factor in politics. No political party can be oblivious of the presence or otherwise of numerically strong caste-groups in a certain region. This information is crucially important in chalking out the electoral strategy at a given point of time. In the absence of updated data on caste no social science debate on the role of caste in electoral politics, economic development and education can be meaningful.

It may be interesting to examine the geographic patterns of distribution of different caste-groups in British India. As is generally known, the last point of time when comprehensive data on caste was collected was during 1931 census, almost 70 years from now. The 1941 census was sketchy due to the impact of war on census operations. The 1931 census of India collected data for all the caste-groups living in the country in the British districts as well as the princely states, adopting not a very satisfactory scheme of classification, but nonetheless trying to capture the prevailing social reality of caste identities.

The massive data generated during the census operations were published along with the reports of the 1931 census under various heads. For example, tables were titled as "territorial distribution of principal communities" (Madras presidency), "caste, tribe or race" (Assam), "race, tribe or caste" (United Provinces of Agra and Oudh), and so on. Population was classified into Brahmanic Hindu, non-Brahmanic Hindu and Muslim castes. Tribal groups were mixed up with castes.

They were further cross-classified by religious persuasions. Using the district-level data of 1931 census, one can broadly identify the geographic patterns in the distribution of different caste-groups. This will perhaps serve as a benchmark. It may be noted that certain caste-groups have been very mobile in space. Moreover, the partition of the country in 1947 resulted in far-reaching changes in population composition in different parts of India. Urbanization, migration for employment, and industrialization, among other things, have contributed to this spatial mobility. This study is, therefore, no more than of historic significance. Admittedly, a comprehensive study of 1931 data on caste-groups calls for an independent exercise at the all-India scale, and may not be feasible within the scope of this work.

This study is, therefore, by choice restricted to the three British provinces, *viz.*, United Provinces of Agra and Oudh, Madras Presidency and Assam, as in 1931. However, the political map of India has also changed since then. While the United Provinces of Agra and Oudh, now known as Uttar Pradesh, have not registered any change in the outer boundary, the Madras Presidency and Assam have changed drastically. The British presidency of Madras incorporated parts of southern Orissa, coastal Andhra Pradesh, Tamil Nadu and parts of Kerala and Karnataka. Similarly, several states have been carved out of the British province of Assam.

United Provinces of Agra and Oudh : The British province consisted of 48 districts directly administered by the British and three princely states of Rampur, Tehri-Garhwal and Benaras. These princely states have since then been merged with Uttar Pradesh. Several British districts have also been bifurcated. The other significant change was the merger of the several tiny princely states of the erstwhile Central India Agency with the Bundelkhand districts of the present state of Uttar Pradesh. Despite these politico-administrative changes within the state, the outer boundary has remained by and large unchanged.

The Stratification

Castes are often classified into three main groups:

(a) functional or occupational castes,

(b) racial or ethnic castes, and

(c) sectarian castes.

Included among the occupational category are the priestly caste of Brahmans, and trading castes, such as the Baniya or Vaishya, Khattri of Punjab, Aggarwal and Oswal of Rajasthan, etc. In fact, there are innumerable caste-groups belonging to the category of occupational castes. They include the whole range of artisans from weavers and carpenters to potters, goldsmiths, barbers, washermen, and so on. Also included in this category are astrologers, oilmen, cattle-breeders, musicians, folk-dancers, acrobats, snake-charmers, and so on. The racial or ethnic castes vary regionally. There are, for example, Rajbansis and Chandals in Bengal and Bihar, Rajbhar and Chero in Uttar Pradesh and Bihar; Jats, Gujjars and Meos in Punjab, Haryana and Rajasthan; Koli and Mahar in Maharashtra; Nair (Nayar) in Kerala; and Paraiyan in Tamil Nadu and Kerala.

Lingayats appear to be the best example of a sectarian caste. In fact, they originated from a religious sect. In Tamil Nadu, Andhra Pradesh and Karnataka, the different groups of Brahmans, such as Telugu Brahmans, Kanarese Brahmans and Oriya Brahmans, reflect the differences perceived in language, cultural tradition and regional identity. Tamil Visvabrahmans and Telugu Visvabrahmans are other examples of regional variations in caste identity. This confirms the view that caste is relevant within a language domain only. It is in the linguistic context that the origins of a particular lineage group can be traced. The data on caste-groups in the United Provinces of Agra and Oudh show a vast variety. Subsumed within the castes are occupational groups, ethnic groups and others with distinctly different origins. No distinction has been made between castes and tribes. Thus, the list of castes also includes several tribes such as the Bhil, Bhoksa, Gond and the Tharu.

The Grouping

The 1931 census enumerated 99 castes under the Brahmanic Hindu category in the United Provinces of Agra and Oudh.

Besides, there were other categories such as '*no caste*' and '*caste unspecified*. The total population of Brahmanic Hindu castes was around 41.5 million persons. In order to understand the pattern of distribution 99 caste-groups have been classified into major, medium, minor and insignificant categories (Table). It may be noted that there were 10 caste-groups in the major category with populations ranging between 1 and 6 million.

Their cumulative strength was about 25.8 millions. There were 46 caste-groups in the medium category with population ranging between 50,000 and 1 million, their cumulative population being 14.5 millions. Under the minor category were 25 castes with a population ranging between 1,000 and 50,000. Their total strength was 0.48 million only. There were 18 castes in the insignificant category with population ranging between 1 and 1,000 persons. Their cumulative population was just 5,000. Moreover, a total population of about 5,000 could not declare their caste, while the caste of another 2,000 persons remained unspecified. The major castes accounted for 62.3 per cent of the Brahmanic Hindu population. The castes of the medium category constituted about 35 per cent of the population, while the share of the minor castes was as small as 1 per cent (Table).

United Provinces of Agra and Oudh: Numerical Ordering of Caste Groups, 1931

Category	*Range of Population*	*Number of Caste groups*	*Total Population*	*Percentage to Total of Brahmanic Hindus*
Major	1,000,000 +	10	25,880,042	62.31
Medium	50,000-1,000,000	46	14,495,448	34.90
Minor	1,000-50,000	25	478,685	1.15
Insignificant	1-1,000	18	5,040	0.012
No Caste	—	—	4,698	0.011
Caste Unspecified	—	—	1,919	0.005

Included in the category of major castes were Chamar, Brahman, Ahir, Rajput, Kurmi, Pasi, Vaishya, Kahar, Lodh and Gadariya groups. The medium category included some well-known castes such as Jats, Gujjars and Kayasthas as well as common occupational castes, such as Kumhar (potter), Lohar (blacksmith), Teli (oilman), Mehtar (scavenger), Mallah and Kewat (boatman), Sonar (goldsmith), Barahi (carpenter), Baghban (gardner), Darzi (tailor), Khatik (butcher) and Halwai (confectioner).

The category of minor castes included occupational groups, such as Mochi (cobbler), Faqir (beggar), Nut (acrobat), Beldar (construction labourer) and Bahelia (bird trapper). Included in this category were also castes, like Khatri, Khangar and Saharia.

Mode of Distribution

In order to understand the patterns of spatial distribution 17 caste-groups have been selected. Their population as well as percentage share are given in Table. It is evident that eight caste-groups were numerically very strong as their population exceeded 1 million persons each. There were nine caste-groups among the 17 castes under study whose population ranged between 1 lakh and 7.5 lakhs.

Together these selected castes accounted for 66.68 per cent of the Brahmanic Hindu population. The selected castes include occupational groups such as Brahman, Vaishya, Teli, Nai, Mehtar, Mallah and Khatik. They also include ethnic castes such as Rajput, Jat, Gujjar, Ahir and Dom. In terms of numerical strength, Chamars were holding the first rank. They outnumbered all other castes, Brahmans with a population of 4.5 million followed in numerical strength.

United Provinces of Agra and Oudh: Population of Selected Brahmanic Castes, 1931

Caste	*Population*	*Per cent of All Brahmanic Hindus*
Chamar	6,292,338	15.15
Brahman	4,525,893	10.90
Ahir	3,886,045	9.36
Rajput	3,538,054	8.52
Kurmi	1,750,557	4.21
Pasi	1,459,940	3.51
Vaisya	1,169,914	2.82
Lodh	1,095,793	2.64
Teli	751,648	1.81
Jat	712,803	1.72
Nai	659,430	1.59
Mehtar	482,718	1.16
Kayastha	467,088	1.12
Mallah	295,673	0.71
Gujjar	287,615	0.69
Khatik	209,668	0.50
Dom	109,906	0.26
Total Population	27,695,083	66.68
All Brahmanic Hindus	41,534,981	
Total Hindu Population	41,856,310	

Notes: Castes arranged in descending order by numerical strength.

Source: Census of India, 1931, Vol. XVIII, United Provinces of Agra and Oudh,

Part II - *Imperial and Provincial Tables*, Table XVII, *Race, Tribe or Caste*, pp. 500-19.

It may be noted that many of these caste-groups are ubiquitous (*e.g.*, Chamar, Brahman, Vaishya, Teli, Nai, Mehtar and Khatik). Similarly, Ahirs and Rajputs are fairly widespread, although they display a tendency of concentration in a few districts. On the other hand, castes such as Gujjar, Jat, Lodh, Kurmi and Dom display a high degree of concentration in certain parts of the province.

Chamar : At the provincial level Chamars constituted 15 per cent of the Brahmanic Hindu population. However, there were several districts in which their proportion was much higher than the provincial average. Notable among these districts were Saharanpur (33.5 per cent), Bijnor (27.7 per cent), Moradabad (23.3 per cent), Muzaffarnagar (22.6 per cent) and Azamgarh (22.3 per cent). On the other hand, there were districts in the Kumaon division in which their share was less than 1 per cent, Nainital being an exception. Percentages were also low in Bahraich (9.3 per cent), Pilibhit (8.2 per cent) and Gonda (3.9 per cent).

Brahman : Brahmans constituted some 11 per cent of population at the provincial level. However, there were several districts in which Brahmans were more preponderant. For example, in the hilly districts of United Provinces, such as Nainital, Almora and Garhwal as well as the princely state of Tehri-Garhwal their proportion ranged between 20 and 24 per cent. Likewise, the proportion of Brahmans was larger in Muttra and Cawnpore—the respective percentages being 17.6 and 15.2.

The districts where the proportionate share of Brahmans was much lower than the provincial average included Muzaffarnagar and Etah (both 6.6 per cent), Mainpuri (6.1 per cent), Saharanpur (5.6 per cent), Moradabad (5.3 per cent), Bijnor (4.5 per cent) and Rampur (4.2 per cent). Understandably, the proportionate share of Brahmans among the Hindus was low in districts with high Muslim population.

Ahir : Ahirs accounted for 9.4 per cent of the Brahmanic Hindu population of the United Provinces. They were largely an insignificant component of population in the Meerut as well as Rohilkhand divisions. Likewise, their proportion was generally low in Muttra and Agra. They, however, displayed a tendency of clustering in many districts of Allahabad division, their proportionate share being far above the provincial average. Ahirs accounted for about 20 per cent of Brahmanic Hindu population in Azamgarh and 18 per cent in Jaunpur. On the other hand, in Kumaon division, they were conspicuous by their absence.

Rajput : Rajputs were fairly widespread in the United Provinces. However, their proportionate share remained around 6 to 8 per cent, the provincial average being 8.5 per cent. It is interesting to note that in the hilly districts of United Provinces Rajputs constituted a high proportion of the Brahmanic Hindu population. In Garhwal, for example, their proportion was as high as 56.5 per cent. Similarly, in the princely state of Tehri-Garhwal, their proportion was even higher, *i.e.*, 59.9 per cent. The other hilly districts of Almora, Dehra Dun and Nainital also had large proportions of Rajputs, the respective percentages being 49.7, 29.2 and 27.5. This strong presence of Rajputs in the hilly districts was an outcome of migration from the southern plains after the Rajputs lost power in the plains during the medieval period. There were two districts with high Rajput percentages in the plains of United Provinces. These were: Bijnor in the west and Balia in the east, the respective percentages being 13.5 and 15.7.

Vaishya : Vaishyas have always been an important component of population. Being the trading community they handled almost exclusively all exchange of goods and offered credit facilities to the people. But their proportionate share was always low. At the 1931 census, the provincial average percentage of Vaishyas was just 2.8. There were not many districts in which the percentage of Vaishyas marked any significant departure from this average. Reference may, however, be made to the western districts of Muzaffarnagar, Meerut, Agra and Muttra where their proportionate share in population was nearly 5 to 6 per cent. The reasons for this relatively high proportionate share may be historical.

Kurmi : With a provincial average of 4.2 per cent Kurmis were not widely distributed. Percentages were very high in Bareilly (14.5), Bara Banki (14.2) and Pilibhit (12.1). Kurmis constituted a large proportion of population in Benares and Partapgarh (10.6 per cent each), as well as Basti (8.7 per cent) and Mirzapur (8.5 per cent). Their percentages were also generally high in the districts of Faizabad division. The case of Bara Banki has already been referred to.

Pasi : As a caste Pasis constituted an insignificant proportion of the Brahamanic Hindu population of the United Provinces. The provincial average was as low as 3.5 per cent. They were fairly preponderant in the districts of Lucknow division with the percentages ranging between 11 and 17. Their proportionate share was also high in Bara Banki. In Allahabad the proportionate share of Pasis was around 10 per cent. It was about 9 per cent in Partapgarh.

Bibliography

Arvill, R. : *Practical Geography*, Penguin, New York, 1967.

Barton, Jonathan R. : *Human Geography of the World*, Routledge, New York, 1997.

Benevolo, L. : *The History of the Human Geography*, Cambridge, London, 1980.

Berry, Brian, J.L. : *The Human Consequences of Urbanization*, St. Martin's Press, New York, 2000.

Bose, A. : *India's Urbanization*, Kanishka Pub., New Delhi, 1980.

Bose, N.K. : *Tribal Life in India*, National Book Trust, New Delhi, 1971.

Brush, John E. : *The Morphology of Indian Cities*, Popular Prakashan, Bombay, 1962.

Burger, J. : *First Peoples*, Robertson McCarta Ltd., London, 1990.

Chandna, R.C. : *A Geography of Rural Population*, Kalyani Publishers, Ludhiana, 1992.

Chapman, G.P. and K.M. Baker : *The Changing Geography of Asia*, Routledge, New York, 1992.

Chard, C.S. : *Man in Prehistory*, McGraw Hill, London, 1969.

Chauhan, V.S. and Gautam, A. : *Comprehensive Geography of India*, Rastogi Publications, Meerut, 1996-97.

Chisholm, M. : *Modern World Development : A Geographical Perspective*, Barnes & Noble, New Jersey, 1982.

Chisholm, Michael : *Rural Settlement : An Essay in Location*, Hutchinson, London, 1966.

Clarke, J.I. : *Geography and Population : Approaches and Applications*, Pergamon Press, New York, 1984.

Clarke, J.I. : *Human Geography*, Pergamon Press, Oxford, 1972.

De Blij, H.J. & Murphy, Alexander, B. : *Human Geography : Culture, Society and Space*, John Wiley and Sons, New York, 1999.

De Blij, H.J. : *The Earth : An Introduction to Its Physical and Human Geography*, John Wiley & Sons, New York, 1995.

Delgado, C. M. : *Readings in Cultural Geography*, Chicago University Press, Chicago, 1962.

Dubey, R.N. : *Human Geography of India*, Kitab Mahal, Allahabad, 1998.

Dutt, A.K.G. Pomeroy and V. Madhwa : *Cultural Patterns of India*, Indian Geography Foundation, Allahabad, 1996.

Farmer, B.H. : *An Introduction to Human Geography*, Routledge, New York, 1993.

Fellman9n, J. : *Human Geography : Landscape of Human Activities*, WC Brown, Dubuque, 1990.

Gopalakrishnan, Ramamoorthy : *Geography of India*, Jawahar Publishers, New Delhi, 1996.

Gorsline, Marie and Gorsoline, D. : *North American Indians*, Random House, New York, 1978.

Goudie, A. : *The Human Impact on the Natural Environment*, Basil Blackwell, Oxford, 1986.

Hammond, Charles W. : *Elements of Human Geography*, Unwin Hyman, London, 1985.

Haq, M.U. : *Human Development in Changing World*, Oxford, New York, 1992,

Jackson, P. and Smith, S.J. : *Exploring Social Geography*, Allen & Unwin, London, 1984.

Jacobson, J.L. : *Planning the Global Family*, World Watch Institute, Washington, 1987.

Jordan, T.G. : *The European Culture Area : A Systematic Geography*, Harper Collins, New York, 1996.

Kothari C.R. : *Research Methodology : Methods and Techniques*, Wiley Eastern, New York, 1990.

Lebon, J.H. : *An Introduction to Human Geography*, Hutchinson, London, 1969.

Leong, G.C. and Morgan, G.C. : *Human and Economic Geography*, Oxford University Press, Oxford, 1992.

Lewin, K. : *Field Theory in Social Science*, University of Chicago Press, Chicago, 1951.

Lewis, G.J. : *Human Migration*, MacMillan, London, 1982.

Misra, R.P. : *Human Geography : Concept, Technique and Policies*, Concept, New Delhi, 1992.

Muller, S.H. : *The World's Living Languages*, Oxford, New York, 1964.

Muthiah, S. : *An Atlas of India*, Oxford University Press, New York, 1992.

Nielsen, N.C. : *Religions of the World*, Oxford, New York, 1983.

Northam, Ray : *Urban Geography*, Wiley, New York, 1975.

Perpillou, A.M. : *Human Geography*, Longman, New York, 1977.

Qureshi, M.H. : *India : Resources and Regional Development*, N. C. E. R. T., New Delhi, 1990.

Rapp, A. : *Human Activity and Environmental Processes*, John Wiley, Chichester, 1987.

Raza, M. : *Development and Ecology*, Rawat Publications, Jaipur, 1992.

Scientific American : *The Human Geography*, W.H. Freeman, San Francisco, 1974.

Short, J.R. : *Introduction to Urban Geography*, Routledge & Kegan Paul, London, 1984.

Singh, L.R. : *A Study in Human Geography*, Ram Narain Beni Prasad, Allahabad, 1965.

Smith, D.M. : *Human Geography : A Welfare Approach*, Edward Arnold, London, 1977.

Tarrant, J.R. : *Human Geography*, Halstead Press, New York 1974.

Van Eckelen, W.F. : *Indian Foreign Policy and the Border Dispute*, Hague, London, 1967.

White, Paul : *The West European City : A Social Geography*, Longman, London, 1984.

Wilfried, B. : *Atmospheric Pollution*, McGraw Hill, New York, 1972.

Yantis : *The Encyclopaedia of World Cultures*, MacMillan, New York, 1993.

Zelinsky, W. : *A Prologue to Population Geography*, Prentice Hall, Englewood Cliffs, 1966.

Index

B

C

M

N

O

P

Q

R

U

V

W

Z

❑❑❑